# The Best American Science and Nature Writing 2004

# The Best American Science and Nature Writing 2004

*Edited and with an Introduction by* Steven Pinker

Tim Folger, Series Editor

HOUGHTON MIFFLIN COMPANY

BOSTON · NEW YORK 2004

ISSN 1530-1508
ISBN 0-618-24697-5
ISBN 0-618-24698-3 (pbk.)

Printed in the United States of America

MP 10 9 8 7 6

"Genesis of Suicide Terrorism" by Scott Atran. First published in *Science*, March 7, 2003. Copyright © 2003 by AAAS. Reprinted by permission.

"The Battle for Your Brain" by Ronald Bailey. First published in *Reason*, February 2003. Copyright © 2003 by Reason Foundation. Reprinted by permission of Mike Alissi, Publisher.

"Fearing the Worst Should Anyone Produce a Clone" by Philip M. Boffey. First published in *The New York Times*, January 5, 2003. Copyright © 2003 by the New York Times Co. Reprinted with permission.

"The Bittersweet Science" by Austin Bunn. First published in *The New York Times Magazine*, March 16, 2003. Copyright © 2003 by Austin Bunn. Reprinted by permission of *The New York Times Magazine*.

"The New Celebrity" by Jennet Conant. First published in *Seed*, February 2003. Copyright © 2003 by Jennet Conant. Reprinted by permission of *Seed*.

"The Mythical Threat of Genetic Determinism" by Daniel C. Dennett. First published in *The Chronicle of Higher Education*, January 31, 2003. Copyright © 2003 by Daniel C. Dennett. Reprinted by permission of the author.

"We're All Gonna Die!" by Gregg Easterbrook. First published in *Wired*, July 2003. Copyright © 2003 Condé Nast Publications. All rights reserved. Reprinted by permission.

"Far-Out Television" by Garrett G. Fagan. First published in *Archaeology*, July/August 2003. Copyright © 2003 by *Archaeology*. Reprinted by permission.

"A War on Obesity, Not the Obese" by Jeffrey M. Friedman. First published in *Science*, February 7, 2003. Copyright © 2003 by AAAS. Reprinted by permission.

# Contents

# Foreword

COMMON SENSE is much overrated as a virtue. The defenders of a flat Earth invoked it. (And a bit of Googling shows that some still do.) No doubt Galileo's inquisitors appealed to it as well. So did critics of the Wright brothers, right up to that first flight just over a century ago. When it comes to understanding the universe, the history of science shows common sense to be a most unreliable guide.

What, for example, would common sense make of the cosmologist Max Tegmark's musings about parallel universes, a story that begins on page 175? "The simplest and most popular cosmological model today," Tegmark writes, "predicts that you have a twin in a galaxy about 10 to the $10^{28}$ meters from here." And that's just the beginning. If the universe is infinite in size, Tegmark adds, it contains infinitely many different versions of you, with each one living every possible variation of your life. One of you is reading this right now. A more fortunate doppelgänger is vacationing in a Tuscany on a mirror Earth countless light-years away from us.

Granted, common sense has its place. Evolution and natural selection have equipped most of us with enough smarts to avoid rash actions that might adversely affect our survival. ("Better not kick the motorcycle of the hulking biker with the swastika tattoo on his forehead, even if he did give me a dirty look.") But unfortunately, evolution hasn't adequately prepared us to easily understand things like quantum mechanics, curved space-time, or even the age of Earth, let alone that of the universe. In those spheres common sense fails completely. Metaphor helps, but it's really just a crutch, emphasizing more often than not the limits of our imagination when confronted with the overwhelming vastness of the cosmos.

Consider how the late Carl Sagan once tried to give some feel for the immensity of the 4.5 billion years that our planet has existed. Imagine allotting yourself 4.5 billion years for a walking tour from San Francisco to New York. On that trip, Sagan calculated, you would take one step every 2,000 years. With numbers like that, it's not hard to see why some people find comfort in the cozier cosmos expressly fitted for us by religion, where the heavens are close at hand and where Earth's creation lies within a span commensurate with human history and comprehension.

But if we don't want to fool ourselves or be lulled into complacency by myths, then to understand the universe — and our place in it — we need a healthy exposure to the *un*common sense of science. Given the sorry track record of the commonsense approach to knowledge, common sense itself suggests that its advocates should have been routed long ago. Sadly, that's not the case. (Chalk up another failure for common sense.) Education officials in Georgia last year, acting with unenviable common sense, proposed to remove references to Earth's age and the word "evolution" from all textbooks used in the state's public schools. To paraphrase the great physicist Wolfgang Pauli, the arguments of the antievolution crowd are so bad they're not even wrong. Those who doubt the reality of evolution must also not believe in the existence of antibiotic-resistant bacteria, a graphic and potentially lethal example of evolution in action, not over millennia but over months and years. As for Earth's age, the very same laws that govern radioactive decay, which provides a tick-by-tick measure of time eon after eon, also describe the flow of electrons in cell phones, computers, and televisions. But the creationists, the advocates of "intelligent design" as an alternative to evolution, the enemies of rationality, refuse to see technology for what it is — an integral part of science, a by-product of our understanding of how the universe works, an understanding that leaves no room for divine intervention or a 6,000-year-old Earth. They apparently place a blind faith in technology — flying on planes, watching satellite TV, downloading Web pages — caring not at all how any of it works.

Reality is far more intriguing than the defenders of dogma dare to imagine. After all, what a dull universe it would be if everything in it conformed to our expectations, if it held nothing to surprise or baffle us or confound our common sense. A century ago no one foresaw the existence of black holes, an expanding universe, oceans on Jupiter's moons, or DNA. What could be more enriching than to know that we share a common origin with all living things, that we are kin to chimpanzees, redwoods, and mollusks? And isn't it a source of wonder to realize that the iron in our blood and the calcium in our bones were created in the bellies of supernovas?

Just a few minutes ago I watched a NASA announcement on CNN that the *Opportunity* rover, now puttering about in a Martian crater, confirmed that water — and perhaps even salty seas — once existed on Mars. The discovery dramatically increases the odds that Mars may have supported life in the past, and perhaps still does. If evidence of life is ever found elsewhere, will some benighted officials, acting in the name of common sense and revealed truth, insist on censoring all mention of it in their schools' books?

It gives me pause to realize that some of the stories collected here could be banned in some schools. Horace Freeland Judson's "The Stuff of Genes" mentions the "E word" a few times in an elegant essay celebrating the fiftieth anniversary of the discovery of DNA, a discovery with ramifications that are just beginning to unfold. Jennet Conant's lively profile of James Watson, the codiscoverer of DNA with Francis Crick, wouldn't pass muster either, given Watson's disdain for religion and his pugnacious zest for controversy. Nicholas Wade's entertaining article, "A Prolific Genghis Khan, It Seems, Helped People the World," would elicit frowns from dour censors with its discussion of human genes. But I think you'll enjoy learning that 16 million men alive today descend from the Mongol warlord. Meredith Small, I'm afraid, would offend those who abhor any hint of a human-ape connection with her meditation on the ways of snow monkeys, in "Captivated." Most of the other articles don't touch at all on evolution, but all have one thing in common. They are uncommonly good.

One of the perks of being the series editor of this book is the opportunity to work with some of the world's most original thinkers. Steven Pinker, this year's guest editor and a contributor last year, has put together a remarkably diverse selection, which he introduces more fully in the next few pages. Deanne Urmy and Melissa Grella at Houghton Mifflin have, as always, with good humor and common sense (of the best sort), guided this book to completion and reined me in when necessary. And the beauteous Anne Nolan, against all dictates of common sense, has chosen to cast her lot with me, a decision for which I am daily grateful.

TIM FOLGER

# Introduction

HORACE'S SUMMARY of the purpose of literature, "to delight and instruct," is also not a bad summary of the purpose of science and nature writing. The difference is not so much that a science essay gives more weight to the second infinitive as that it unites the two. The best science writing delights *by* instructing. A good science essay, like any good essay, must be written with structure and style, but the best science essays accomplish something else. They give readers the blissful click, the satisfying aha!, of seeing a puzzling phenomenon explained.

A good example of what I have in mind comes from my days as a graduate student. Not from an experience in graduate *school* but from an experience living in the kind of apartment that graduate students can afford. One day its antiquated plumbing sprang a leak, and an articulate plumber (perhaps an underemployed Ph.D., I feared) explained what caused it. Water obeys Newton's second law. Water is dense. Water is incompressible. When you shut off a tap, a large incompressible mass moving at high speed has to decelerate very quickly. This imparts a substantial force to the pipes, like a car slamming into a wall, which eventually damages the threads and causes a leak. To deal with this problem, plumbers used to install a closed vertical section of pipe, a "pipe riser," near each faucet. When the faucet is shut, the water compresses the column of air in the riser, which acts like a shock absorber. Unfortunately, Henry's Law applies: Gas under pressure is absorbed by a liquid. Over time, the air in the column dissolves into the water, which fills the pipe riser, rendering it useless. So every now and then a plumber has to bleed the system and let air back into the risers, a bit of preventive maintenance the landlord had neglected. It may not be the harmony of the spheres or the grandeur in this view of life, but the plumber's disquisition captured what I treasure most in science writing: the ability to show how a seemingly capricious occurrence falls out of laws of greater generality.

Good science writing has to be good writing, and another graduate school experience led me to appreciate its first priority, clarity. The great Harvard psychologist Gordon Allport had died years before I entered the program, but he had written an "Epistle to Thesis Writers" that was still being handed down from generation to generation of doctoral candidates. Allport tried to steer students away from the clutter and fog of professional science prose and offered as a model an essay by a ten-year-old girl, who, he wrote, merited a higher degree "if not for the accuracy of her knowledge, then at least for the clarity of her diction":

> The bird that I am going to write about is the Owl. The Owl cannot see at all by day and at night is as blind as a bat.
>
> I do not know much about the Owl, so I will go on to the beast I am going to choose. It is the Cow. The Cow is a mammal. It has six sides — right, left, an upper and below. At the back it has a tail on which hangs a brush. With this it sends the flies away so that they do not fall into the milk. The head is for the purpose of growing horns and so that the mouth can be somewhere. The horns are to butt with, and the mouth is to moo with. Under the cow hangs the milk. It is arranged for milking. When people milk, the milk comes through and there is never any end to the supply. How the cow does it I have not yet realized, but it makes more and more. The cow has a fine sense of smell; one can smell it far away. This is the reason for the fresh air in the country.
>
> The man cow is called an ox. It is not a mammal. The cow does not eat much, but what it eats it eats twice, so that it gets enough. When it is hungry it moos, and when it says nothing it is because its inside is all full up with grass.

In assembling this collection I looked for essays that combined the explanatory depth of the plumber with the limpid prose of the young zoologist. Explanatory depth, surprisingly, is not that easy to find. The most common specimen is the science news story. A journalist flips through the contents of *Science, Nature,* and the *New England Journal of Medicine,* finds the article with the weirdest or most alarming or most bite-sized finding, gets a quote from an author, a supporter, and a critic, and reports that the discovery has overturned everything that scientists had always believed. I understand the pressures that shape this formula: the drama of iconoclasm, the demand by editors for news rather than pedagogy, a desire to show that science is a human activity among spirited antagonists rather than a revelation of the truth by white-coated priests.

But just as presidential campaigns can be distorted by the press's obsession with minute-by-minute changes in popularity polls, an understanding of science can be poorly served by news from the front about continual revolution. Conclusions from individual experiments, especially the most surprising ones, are more ephemeral than conclusions from the reviews and syntheses that can't be squeezed into a brief report in *Science.* The discovery-du-jour approach can whipsaw readers between contradictory claims of uneven worth or leave them with lasting misimpressions, such as that everything that is pleasurable is deadly for one reason or another. And contrary to the idea (commonly associated with Karl Popper) that science is a kind of skeet shooting whose goal is to put a bullet through one hypothesis after another, the best science weaves observations into an explanatory narrative. "All the Old Sciences Have Starring Roles" by Chet Raymo (whose weekly science column graced the *Boston Globe* until his retirement this year) makes the point succinctly. Max Tegmark's mind-expanding "Parallel Universes" shows, by example and argument, how a powerful theory can not only organize sundry data but also lead to an exhilarating new conception of reality itself. Horace Judson's "The Stuff of Genes" reflects on the far-flung implications — for science and life — of the discovery of the structure of DNA, whose golden jubilee was marked in the year these essays appeared.

Clarity and style, happily, are not in short supply in today's science writing (though in professional journals their frequency is commensurate with galliformes' dentition). The genre continues to attract fine writers of all ages, belying the plaint that the younger generation no longer cares about language. Of all the things that go into good science writing, I am fondest of prose that airs out a stuffy hall of scholarship and conveys its insights (or its absurdities) with irreverent wit, like Gregg Easterbrook's "We're All Gonna Die!," Jonathan Rauch's "Caring for Your Introvert," Ron Rosenbaum's "Sex Week at Yale," and Robert Sapolsky's "Bugs in the Brain." But pride of place goes to the Bird Folks (the nom de plume of Mike O'Connor) at the Bird Watcher's General Store in Orleans, Massachusetts, whose informative weekly column strikes a tone that is opposite to the worshipful sonorities found in much nature writing (parodied by Mark Twain as "Far in the empty sky a solitary esophagus slept upon motionless wing").

"Ask the Bird Folks" could have seen the light of day only in a quirky rural tabloid, and I think the proliferation of other unconventional outlets will be a boon to unorthodox styles, formats, and, most important, opinions. Conventional wisdom can jell prematurely when a few commentators stake out the cramped real estate in national publications, and I actually believe the old cliché that the Internet is changing intellectual life by providing limitless outlets for unconventional ideas. As it happens, most of the Web pieces on my short list were dropped at the last minute because of various exigencies (wrong year, wrong country, too much overlap). I suspect that more and more of our best science writing will be found on sites like www.edge.org, scitechdaily.com (and its sister site artsandlettersdaily.com), www.spiked-online.com, butterfliesandwheels.com, techcentralstation.com, human-nature.com, and the many blogs by science-oriented journalists.

Science *is* a human activity, of course, and its rewards are not just discovery and explanation. Most scientists enjoy the mundane activity of gathering their kind of data, and in "Captivated" Meredith Small shares with her readers the pleasures of primatology (while making me understand for the first time why primates groom). And the passionate eccentrics who call themselves scientists are good grist for gossip and character studies, such as Jennet Conant's profile, which presents yet another consequence of the DNA revolution: the appearance on the scientific stage of the inimitable James Watson.

Perhaps more than is usual in these collections, my choices are slanted toward human behavior, and their methods shade into the social sciences. In part this reflects my own interests in psychology, linguistics, neuroscience, and evolution. It may also show that human interest makes for the most compelling writing. But most of all, it reflects the fact that the study of the mind will be among the liveliest frontiers of science in the coming century.

One of these frontiers is the application of genomic analyses to the mind and its products, often in highly unpredictable ways. Judson alludes to the recent finding that the normal version of a gene for a speech and language disorder bears the statistical fingerprints of natural selection acting in our lineage after it split off from the lineage leading to chimpanzees. In one stroke this discovery obliterates the suspicion that the evolution of language and mind is permanently beyond the reach of rigorous science. The duo by Nicholas Wade, "In Click Languages, an Echo of the Tongues of the Ancients" and "A Prolific Genghis Khan, It Seems, Helped People the World," explain two other remarkable applications of genomics to human evolution. One confirms the idea that aggressive polygyny could have affected human evolution by altering our species' genetic makeup (with a surprise appearance by one of the great villains of history). The other may shed light on what was thought to be forever unknowable: the first language spoken by our species.

In anticipating a steady turning of science to the mind and its products I am thinking not just of fancy technologies but of an extension to human affairs of the scientific mindset itself. This does not mean reducing the human condition to genes or neurons or primate behavior, but rather seeking to ascertain whether a claim about human affairs is consistent with the facts and with everything else we know about how the world works. Today this attitude is far from universal. What would happen if newspapers imposed the following rule: any pundit who comments on a trend and blames it on some factor must adduce evidence that (a) the trend is real, (b) the factor preceded the trend, and (c) that kind of factor causes that kind of trend? On many days the op-ed page would consist of a vast empty space op the ed.

Many of my choices upend some bit of conventional wisdom about human life. In "The Bloody Crossroads of Grammar and Politics," Geoffrey Nunberg uses a smidgen of linguistics to expose a bit of nonsense about "correct grammar" and the decline of standards that had been latched on to by writers from David Skinner in the *Weekly Standard* to Louis Menand in *The New Yorker.* In "Where Have All the Lisas Gone?" Peggy Orenstein shows that trends in baby names are not inspired by the latest celebrities, the popularity of religion, or just about any other external cause. Virginia Postrel's "The Design of Your Life" presents a sample of the many myths about aesthetics that she dispatches in her 2003 book *The Substance of Style,* such as the notion that people seek beauty only when their other needs are met, that styles are foisted upon a passive public by manipulative advertisers, and that economic value resides in practical goods and services. Jeffrey Friedman's "A War on Obesity, Not the Obese" shows that we are not getting as fat as obesity statistics would suggest and that the solution to this health problem does not consist of finding the right people to blame. "Sex Week at Yale" shows that being an academic is no protection against holding ludicrous beliefs about human motives, such as the dogmas about love and sex that are common in the humanities and helping professions.

Many misconceptions about behavior are harmless, but in these dangerous times some could lead to catastrophe. Steve Sailer's "The Cousin Marriage Conundrum" correctly predicts that it would be unwise to try to graft a political system onto a society without understanding how the psychology of kinship and ethnic identification plays out in the local environment. Scott Atran's "Genesis of Suicide Terrorism" debunks the bromide, endorsed by impressive lists of Nobel Prize winners and other right-thinking people in countless signed statements, that the root causes of terrorism are poverty and ignorance. The article is no more comforting to those who analyze suicide terrorism only in moralistic terms and insist that terrorists are crazed fanatics or callous psychopaths. Moral outrage is certainly an appropriate response to any slaying of innocents, and it is worth considering the possibility that the retaliation or preemption inspired by outrage is an effective countermeasure. But moral condemnation is just one technique of behavior modification, and the fact that it feels right is no guarantee that it will work. If our goal is to minimize innocent deaths, we may have to set aside our moral intuitions long enough to try to understand the behavior in terms of cause and effect, and that means studying the beliefs, desires, and social dynamics of terrorist groups. I suspect that people from all over the political spectrum may be disturbed by Atran's amoral analysis, but it is a mode of thought that we may have to get used to if we want to improve human affairs.

The interface between science and morals also motivates my remaining choices. Much science journalism today is hostile to scientists in much the same way that much political journalism in the post-Watergate era is hostile to politicians. Scientists are often depicted as arrogant Fausts or cruel Mengeles or greedy profiteers. One article I rejected, for instance, denounced a research program that succeeded in modifying corn to synthesize pharmaceuticals cheaply despite its promise of vast enhancements to human health and a demonstrably trivial risk to the environment. Halos are awarded only to whistleblowers in ecology or climate science who warn us about the wages of our technological lifestyle. In Europe, left-leaning greens call for a Precautionary Principle in which applications of science should be banned or restricted if there is some chance they will have harmful effects, even in the absence of scientific evidence that they do. If the policy, aptly satirized as "Never do anything for the first time," had been applied in the past, it would have ruled out every new technology from fire to fertilizers to malaria control to oral contraception. In the United States, right-leaning bioethicists see research to improve health and well-being as a promethean grab at immortality and a soul-deadening quest to rob us of the nobility of suffering.

This hostility is a big change from the reception that scientists enjoyed a generation ago. When I was a child, my favorite literary genre was the hagiography of a famous scientist; I was taught that Sabin and Salk were the pride of the Jewish people and Banting and Best the pride of Canada. No doubt we are all better off today with a more skeptical treatment of science, but we have swung too far in the direction of timidity about the applications of science and cynicism about the motives of scientists. Austin Bunn's "The Bittersweet Science" and Atul Gawande's "Desperate Measures" put a human face on uncured illness and remind us why aggressive medical pioneers were once revered: they lessened pain, infirmity, and needless death, the most noble goal of human striving. There is a story waiting to be told on how the moral coloring of science (and other endeavors) in different periods can be distorted by quirks of the human moral sense (a fertile new research topic in psychology). Our neural circuits for morality are overly receptive to the trappings of purity, naturalness, and custom, and they are too easily impressed by gravitas, indignation, conspicuous asceticism, and other advertisements of saintliness that may have scant correlation with actions that make people better off.

Genetics, neuroscience, and evolutionary biology will call into question other moral intuitions. Reams of nonsense have been written about cloning, genes linked with personality, and pharmaceuticals that may enhance mood, concentration, and memory. Some of the non sequiturs are so bizarre that they make me wonder whether the authors have fully assimilated what Francis Crick calls "the astonishing hypothesis" — the idea that all thought and feeling consist of physiological activity in the brain — and instead tacitly believe that human choice and individuality reside in an autonomous soul. Philip Boffey's "Fearing the Worst Should Anyone Produce a Cloned Baby," Daniel Dennett's "The Mythical Threat of Genetic Determinism," and Ronald Bailey's "The Battle for Your Brain" are breaths of cool thinking in these overheated arenas.

I end with an indulgence. One article that particularly drew me in was, of all things, "Through the Eye of an Octopus." What could a cognitive scientist find so interesting about the secret life of cephalopods? It is not just that the piece reveals an astonishing spectacle in the natural world, and it's not just that the protagonist is named Steve. The reasons are twofold, and it is not too much of a stretch to say that they illustrate another of my favorite themes in science writing: the interconnectedness of all knowledge, no matter how remote the disciplines.

My first reason for liking the article is linguistic. In one of Gary Larson's *Far Side* cartoons, a bespectacled octopus at a podium addresses his conspecifics: "Fellow octopi, or octopuses . . . octopi? Dang, it's hard to start a speech with this crowd." Judging from an Internet search, human scientists also go both ways on this issue. But Eric Scigliano consistently refers to his subjects as *octopuses,* and he has the logic of language on his side. The *-us* in *octopus* is not the Latin masculine noun ending of *alumnus* and *fungus,* which is replaced by *-i* in the plural. No, it is part of the Greek word *pous* meaning foot, and turning it into *-pi* makes no sense. Nor could English have imported the Greek plural as an irregular form, as it did with *criterion-criteria* and *stigma-stigmata,* giving us *octopodes.* An *octopus* is the creature that *owns* the enumerated feet, not the assembly of feet itself. The elegant algorithm that computes the properties of complex words (described in my book *Words and Rules*) ensures that these synecdochic compounds have regular plurals, even when they are built around irregular nouns. Hence we refer to several members of the extinct family of cats as *saber-tooths* (not *saber-teeth*). We similarly talk about *lowlifes, still lifes, tenderfoots, flatfoots,* and, in *The Lord of the Rings, Proudfoots.* And by this linguistic logic, we should identify more than one octopus as *octopuses.*

The other reason I liked the article has to do with human evolution. It's lonely to be one of the few species with advanced powers of problem solving, and it's scientifically frustrating too. How can we test ideas about the evolution of intelligence if it happened only once? One way is to find smarter-than-average species from widely separated branches of the tree of life and see what else distinguishes them from their duller cousins. Studies of other smart creatures like dolphins and wolves suggest that group living is one of the traits that sets the stage for the evolution of higher intelligence. But this does not explain why humans are so much smarter than other social species. I have always suspected that the ancestral ape that spawned our lineage must have been dealt a *number* of traits that made higher intelligence worth its metabolic cost. And I speculated in *How the Mind Works* that one of those traits is the possession of hands. Evolution does not reward cerebration for its own sake but only thoughts that can be put to use in adaptive ways, such as manipulating the world to one's advantage. If this idea is right, intelligence increased in our ancestors partly because they were equipped with levers of influence on the world, namely the grippers found at the ends of their two arms. How pleasing to learn that intelligence also evolved in a species that has eight of them.

STEVEN PINKER

*The Best American Science
and Nature Writing 2004*

SCOTT ATRAN

# Genesis of Suicide Terrorism

FROM *Science*

ACCORDING TO THE U.S. Department of State report *Patterns of Global Terrorism 2001,* no single definition of terrorism is universally accepted; however, for purposes of statistical analysis and policy making: "The term 'terrorism' means premeditated, politically motivated violence perpetrated against noncombatant targets by subnational groups or clandestine agents, usually intended to influence an audience." Of course, one side's "terrorists" may well be another side's "freedom fighters." For example, in this definition's sense, the Nazi occupiers of France rightly denounced the "subnational" and "clandestine" French Resistance fighters as terrorists. During the 1980s, the International Court of Justice used the U.S. administration's own definition of terrorism to call for an end to U.S. support for "terrorism" on the part of Nicaraguan Contras opposing peace talks.

For the U.S. Congress, "'act of terrorism' means an activity that — (A) involves a violent act or an act dangerous to human life that is a violation of the criminal laws of the United States or any State, or that would be a criminal violation if committed within the jurisdiction of the United States or of any State; and (B) appears to be intended (i) to intimidate or coerce a civilian population; (ii) to influence the policy of a government by intimidation or coercion; or (iii) to affect the conduct of a government by assassination or kidnapping." When suitable, the definition can be broadened to include states hostile to U.S. policy.

Apparently, two official definitions of terrorism have existed since the early 1980s: that used by the Department of State "for statistical and analytical purposes" and that used by Congress for criminal proceedings. Together, the definitions allow great flexibility in selective application of the concept of terrorism to fluctuating U.S. priorities. The special category of "State-sponsored terrorism" could be invoked to handle some issues, but the highly selective and politically tendentious use of the label "terrorism" would continue all the same. Indeed, there appears to be no principled distinction between "terror" as defined by the U.S. Congress and "counterinsurgency" as allowed in U.S. armed forces manuals.

Rather than attempt to produce a stipulative and all-encompassing definition of terrorism, this article restricts its focus to "suicide terrorism" characterized as follows: the targeted use of self-destructing humans against noncombatant — typically civilian — populations to effect political change. Although a suicide attack aims to physically destroy an initial target, its primary use is typically as a weapon of psychological warfare intended to affect a larger public audience. The primary target is not those actually killed or injured in the attack, but those made to witness it. The enemy's own information media amplify the attack's effects to the larger target population. Through indoctrination and training and under charismatic leaders, self-contained suicide cells canalize disparate religious or political sentiments of individuals into an emotionally bonded group of fictive kin who willfully commit to die spectacularly for one another and for what is perceived as the common good of alleviating the community's onerous political and social realities.

Suicide attack is an ancient practice with a modern history. Its use by the Jewish sect of Zealots *(sicari)* in Roman-occupied Judea and by the Islamic Order of Assassins *(hashashin)* during the early Christian Crusades are legendary examples. The concept of "terror" as systematic use of violence to attain political ends was first codified by Maximilien Robespierre during the French Revolution. He deemed it an "emanation of virtue" that delivers "prompt, severe, and inflexible" justice, as "a consequence of the general principle of democracy applied to our country's most pressing needs." The Reign of Terror, during which the ruling Jacobin faction exterminated thousands of potential enemies, of whatever sex, age, or condition, lasted until Robespierre's fall (July 1794). Similar justification for state-sponsored terror was common to twentieth-century revolutions, as in Russia (Lenin), Cambodia (Pol Pot), and Iran (Khomeini).

Whether subnational (e.g., Russian anarchists) or state-supported (e.g., Japanese kamikaze), suicide attack as a weapon of terror is usually chosen by weaker parties against materially stronger foes when fighting methods of lesser cost seem unlikely to succeed. Choice is often voluntary, but typically under conditions of group pressure and charismatic leadership. Thus, the kamikaze ("divine wind") first used in the battle of the Philippines (November 1944) were young, fairly well educated pilots who understood that pursuing conventional warfare would likely end in defeat. When collectively asked by Adm. Takijiro Onishi to volunteer for "special attack" *(tokkotai)* "transcending life and death," all stepped forward, despite assurances that refusal would carry no shame or punishment. In the Battle of Okinawa (April 1945) some two thousand kamikaze rammed fully fueled fighter planes into more than three hundred ships, killing five thousand Americans in the most costly naval battle in U.S. history. Because of such losses, there was support for using the atomic bomb to end World War II.

The first major contemporary suicide terrorist attack in the Middle East was the December 1981 destruction of the Iraqi embassy in Beirut (27 dead, more than 100 wounded). Its precise authors are still unknown, although it is likely that Ayatollah Khomeini approved its use by parties sponsored by Iranian intelligence. With the assassination of pro-Israeli Lebanese President Bashir Gemayel in September 1982, suicide bombing became a strategic political weapon. Under the pro-Iranian Lebanese Party of God (Hezbollah), this strategy soon achieved geopolitical effect with the October 1983 truck-bomb killing of nearly three hundred American and French servicemen. The United States and France abandoned the multinational force policing Lebanon. By 1985, these attacks arguably led Israel to cede most of the gains made during its 1982 invasion of Lebanon.

In Israel-Palestine, suicide terrorism began in 1993 with attacks by Hezbollah-trained members of the Islamic Resistance Movement (Hamas) and Palestine Islamic Jihad (PIJ) aimed at derailing the Oslo Peace Accords. As early as 1988, however, PIJ founder Fathi Shiqaqi established guidelines for "exceptional" martyrdom operations involving human bombs. He followed Hezbollah in stressing that God extols martyrdom but abhors suicide: "Allah may cause to be known those who believe and may make some of you martyrs, and Allah may purify those who believe and may utterly destroy the disbelievers"; however, "no one can die except by Allah's leave."

The recent radicalization and networking through Al-Qaida of militant Islamic groups from North Africa, Arabia, and Central and Southeast Asia stems from the Soviet-Afghan War (1979–1989). With financial backing from the United States, members of these various groups were provided opportunities to pool and to unify doctrine, aims, training, equipment, and methods, including suicide attack. Through its multifaceted association with regional groups (by way of finance, personnel, and logistics), Al-Qaida aims to realize flexibly its global ambition of destroying Western dominance through local initiatives to expel Western influences. According to *Jane's Intelligence Review:* "All the suicide terrorist groups have support infrastructures in Europe and North America."

Calling the current wave of radical Islam "fundamentalism" (in the sense of "traditionalism") is misleading, approaching an oxymoron. Present-day radicals, whether Shi'ite (Iran, Hezbollah) or Sunni (Taliban, Al-Qaida), are much closer in spirit and action to Europe's post-Renaissance Counter Reformation than to any traditional aspect of Moslem history. The idea of a ruling ecclesiastical authority, a state or national council of clergy, and a religious police devoted to physically rooting out heretics and blasphemers has its clearest historical model in the Holy Inquisition. The idea that religion must struggle to assert control over politics is radically new to Islam.

Recent treatments of Homeland Security research concentrate on how to spend billions to protect sensitive installations from attack. But this last line of defense is probably easiest to breach because of the multitude of vulnerable and likely targets (including discotheques, restaurants, and malls), the abundance of would-be attackers (needing little supervision once embarked on a mission), the relatively low costs of attack (hardware store ingredients, no escape needs), the difficulty of detection (little use of electronics), and the unlikelihood that attackers would divulge sensitive information (being unaware of connections beyond their operational cells). Exhortations to put duct tape on windows may assuage (or incite) fear but will not prevent massive loss of life, and public realization of such paltry defense can undermine trust. Security agencies also attend to prior lines of defense, such as penetrating agent-handling networks of terrorist groups, with only intermittent success.

A first line of defense is to prevent people from becoming terrorists. Here, success appears doubtful should current government and media opinions about why people become human bombs translate into policy. Suicide terrorists often are labeled crazed cowards bent on senseless destruction who thrive in the midst of poverty and ignorance. The obvious course becomes to hunt down terrorists while simultaneously transforming their supporting cultural and economic environment from despair to hope. What research there is, however, indicates that suicide terrorists have no appreciable psychopathology and are at least as educated and economically well off as their surrounding populations.

U.S. President George W. Bush initially branded 9/11 hijackers "evil cowards." For U.S. Senator John Warner, preemptive assaults on terrorists and those supporting terrorism are justified because: "Those who would commit suicide in their assaults on the free world are not rational and are not deterred by rational concepts." In attempting to counter anti-Moslem sentiment, some groups advised their members to respond that "terrorists are extremist maniacs who don't represent Islam at all."

Social psychologists have investigated the "fundamental attribution error," a tendency for people to explain behavior in terms of individual personality traits, even when significant situational factors in the larger society are at work. U.S. government and media characterizations of Middle East suicide bombers as craven homicidal lunatics may suffer from a fundamental attribution error: No instances of religious or political suicide terrorism stem from lone actions of cowering or unstable bombers.

Psychologist Stanley Milgram found that ordinary Americans also readily obey destructive orders under the right circumstances. When told by a "teacher" to administer potentially life-threatening electric shocks to "learners" who fail to memorize word pairs, most comply. Even when subjects stressfully protest as victims plead and scream, use of extreme violence continues — not because of murderous tendencies but from a sense of obligation in situations of authority, no matter how trite. A legitimate hypothesis is that apparently extreme behaviors may be elicited and rendered commonplace by particular historical, political, social, and ideological contexts.

With suicide terrorism, the attributional problem is to understand why nonpathological individuals respond to novel situational factors in numbers sufficient for recruiting organizations to implement policies. In the Middle East, perceived contexts in which suicide bombers and supporters express themselves include a collective sense of historical injustice, political subservience, and social humiliation vis-à-vis global powers and allies, as well as countervailing religious hope. Addressing such perceptions does not entail accepting them as simple reality; however, ignoring the causes of these perceptions risks misidentifying causes and solutions for suicide bombing.

There is also evidence that people tend to believe that their behavior speaks for itself, that they see the world objectively, and that only other people are biased and misconstrue events. Moreover, individuals tend to misperceive differences between group norms as more extreme than they really are. Resulting misunderstandings — encouraged by religious and ideological propaganda — lead antagonistic groups to interpret each other's views of events, such as terrorism/freedom-fighting, as wrong, radical, and/or irrational. Mutual demonization and warfare readily ensue. The problem is to stop this spiral from escalating in opposing camps.

Across our society, there is wide consensus that ridding society of poverty rids it of crime. According to President Bush, "We fight poverty because hope is the answer to terror. . . . We will challenge the poverty and hopelessness and lack of education and failed governments that too often allow conditions that terrorists can seize." At a gathering of Nobel Peace Prize laureates, South Africa's Desmond Tutu and South Korea's Kim Dae Jong opined, "at the bottom of terrorism is poverty"; Elie Wiesel and the Dalai Lama concluded, "education is the way to eliminate terrorism."

Support for this comes from research pioneered by economist

Gary Becker showing that property crimes are predicted by poverty and lack of education. In his incentive-based model, criminals are rational individuals acting on self-interest. Individuals choose illegal activity if rewards exceed probability of detection and incarceration together with expected loss of income from legal activity ("opportunity costs"). Insofar as criminals lack skill and education, as in much blue-collar crime, opportunity costs may be minimal; so crime pays.

Such rational-choice theories based on economic opportunities do not reliably account for some types of violent crimes (domestic homicide, hate killings). These calculations make even less sense for suicide terrorism. Suicide terrorists generally are not lacking in legitimate life opportunities relative to their general population. As a writer in the Arab press emphasizes, if martyrs had nothing to lose, sacrifice would be senseless: "He who commits suicide kills himself for his own benefit, he who commits martyrdom sacrifices himself for the sake of his religion and his nation. . . . The Mujahed is full of hope."

Research by Krueger and Maleckova suggests that education may be uncorrelated, or even positively correlated, with supporting terrorism. In a December 2001 poll of 1,357 West Bank and Gaza Palestinians eighteen years of age or older, those having twelve or more years of schooling supported armed attacks by 68 points, those with up to eleven years of schooling by 63 points, and illiterates by 46 points. Only 40 percent of persons with advanced degrees supported dialogue with Israel versus 53 percent with college degrees and 60 percent with nine years or less of schooling. In a comparison of Hezbollah militants who died in action with a random sample of Lebanese from the same age group and region, militants were less likely to come from poor homes and more likely to have had secondary-school education.

Nevertheless, relative loss of economic or social advantage by educated persons might encourage support for terrorism. In the period leading to the first Intifada (1982–1988), the number of Palestinian men with twelve years or more of schooling more than doubled; those with less schooling increased only 30 percent. This coincided with a sharp increase in unemployment for college graduates relative to high school graduates. Real daily wages of college graduates fell some 30 percent; wages for those with only secondary schooling held steady. Underemployment also seems to be a factor among those recruited to Al-Qaida and its allies from the Arabian peninsula.

Although humiliation and despair may help account for susceptibility to martyrdom in some situations, this is neither a complete explanation nor one applicable to other circumstances. Studies by psychologist Ariel Merari point to the importance of institutions in suicide terrorism. His team interviewed thirty-two of thirty-four bomber families in Palestine/Israel (before 1998), surviving attackers, and captured recruiters. Suicide terrorists apparently span their population's normal distribution in terms of education, socioeconomic status, and personality type (introvert vs. extrovert). Mean age for bombers was early twenties. Almost all were unmarried and expressed religious belief before recruitment (but no more than did the general population).

Except for being young, unattached males, suicide bombers differ from members of violent racist organizations with whom they are often compared. Overall, suicide terrorists exhibit no socially dysfunctional attributes (fatherless, friendless, or jobless) or suicidal symptoms. They do not vent fear of enemies or express "hopelessness" or a sense of "nothing to lose" for lack of life alternatives that would be consistent with economic rationality. Merari attributes primary responsibility for attacks to recruiting organizations, which enlist prospective candidates from this youthful and relatively unattached population. Charismatic trainers then intensely cultivate mutual commitment to die within small cells of three to six members. The final step before a martyrdom operation is a formal social contract, usually in the form of a video testament.

From 1996 to 1999 Nasra Hassan, a Pakistani relief worker, interviewed nearly 250 Palestinian recruiters and trainers, failed suicide bombers, and relatives of deceased bombers. Bombers were men aged eighteen to thirty-eight: "None were uneducated, desperately poor, simple-minded, or depressed. . . . They all seemed to be entirely normal members of their families." Yet "all were deeply religious," believing their actions "sanctioned by the divinely revealed religion of Islam." Leaders of sponsoring organizations complained, "Our biggest problem is the hordes of young men who beat on our doors."

Psychologist Brian Barber surveyed 900 Moslem adolescents during Gaza's first Intifada (1987–1993). Results show high levels of participation in and victimization from violence. For males, 81 percent reported throwing stones, 66 percent suffered physical assault, and 63 percent were shot at (versus 51, 38, and 20 percent for females). Involvement in violence was not strongly correlated with depression or antisocial behavior. Adolescents most involved displayed strong individual pride and social cohesion. This was reflected in activities: For males, 87 percent delivered supplies to activists, 83 percent visited martyred families, and 71 percent tended the wounded (57, 46, and 37 percent for females). A follow-up during the second Intifada (2000–2002) indicates that those still unmarried act in ways considered personally more dangerous but socially more meaningful. Increasingly, many view martyr acts as most meaningful. By summer 2002, 70 to 80 percent of Palestinians endorsed martyr operations.

Previously, recruiters scouted mosques, schools, and refugee camps for candidates deemed susceptible to intense religious indoctrination and logistical training. During the second Intifada, there has been a surfeit of volunteers and increasing involvement of secular organizations (allowing women). The frequency and violence of suicide attacks have escalated (more bombings since February 2002 than from 1993 to 2000); planning has been less painstaking. Despite these changes, there is little to indicate overall change in bomber profiles (mostly unmarried, average socioeconomic status, moderately religious).

In contrast to Palestinians, surveys with a control group of Bosnian Moslem adolescents from the same time period reveal markedly weaker expressions of self-esteem, hope for the future, and prosocial behavior. A key difference is that Palestinians routinely invoke religion to invest personal trauma with proactive social meaning that takes injury as a badge of honor. Bosnian Moslems typically report not considering religious affiliation a significant part of personal or collective identity until seemingly arbitrary violence forced awareness upon them.

Thus, a critical factor determining suicide terrorism behavior is arguably loyalty to intimate cohorts of peers, which recruiting organizations often promote through religious communion. Consider data on thirty-nine recruits to Harkat al-Ansar, a Pakistani-based ally of Al-Qaida. All were unmarried males, most had studied the Quran. All believed that by sacrificing themselves they would help secure the future of their "family" of fictive kin: "Each [martyr] has a special place — among them are brothers, just as there are sons and those even more dear." A Singapore Parliamentary report on thirty-one captured operatives from Jemaah Islamiyah and other Al-Qaida allies in Southeast Asia underscores the pattern: "These men were not ignorant, destitute or disenfranchised. All 31 had received secular education. . . . Like many of their counterparts in militant Islamic organizations in the region, they held normal, respectable jobs. . . . As a group, most of the detainees regarded religion as their most important personal value . . . secrecy over the true knowledge of jihad, helped create a sense of sharing and empowerment vis-à-vis others."

Such sentiments characterize institutional manipulation of emotionally driven commitments that may have emerged under natural selection's influence to refine or override short-term rational calculations that would otherwise preclude achieving goals against long odds. Most typically, such emotionally driven commitments serve as survival mechanisms to inspire action in otherwise paralyzing circumstances, as when a weaker person convincingly menaces a stronger person into thinking twice before attempting to take advantage. In religiously inspired suicide terrorism, however, these emotions are purposely manipulated by organizational leaders, recruiters, and trainers to benefit the organization rather than the individual.

Little tangible benefit (in terms of rational-choice theories) accrues to the suicide bomber, certainly not enough to make the likely gain one of maximized "expected utility." Heightened social recognition occurs only after death, obviating personal material benefit. But for leaders, who almost never consider killing themselves (despite declarations of readiness to die), material benefits more likely outweigh losses in martyrdom operations. Hassan cites one Palestinian official's prescription for a successful mission: "a willing young man . . . nails, gunpowder, a light switch and a short cable, mercury (readily obtainable from thermometers), acetone. . . . The most expensive item is transportation to an Israeli town." The total cost is about $150.

For the sponsoring organization, suicide bombers are expendable assets whose losses generate more assets by expanding public support and pools of potential recruits. Shortly after 9/11, an intelligence survey of educated Saudis (ages twenty-five to forty-one) concluded that 95 percent supported Al-Qaida. In a December 2002 Pew Research Center survey on growing anti-Americanism, only 6 percent of Egyptians viewed America and its "War on Terror" favorably. Money flows from those willing to let others die, easily offsetting operational costs (training, supporting personnel, safe houses, explosives and other arms, transportation, and communication). After a Jerusalem supermarket bombing by an eighteen-year-old Palestinian female, a Saudi telethon raised more than $100 million for "the Al-Quds Intifada."

Massive retaliation further increases people's sense of victimization and readiness to behave according to organizational doctrines and policies structured to take advantage of such feelings. In a poll of 1,179 West Bank and Gaza Palestinians in spring 2002, 66 percent said army operations increased their backing for suicide bombings. By year's end, 73 percent of Lebanese Moslems considered suicide bombings justifiable. This radicalization of opinion increases both demand and supply for martyrdom operations. A December 2002 UN report credited volunteers with swelling a reviving Al-Qaida in forty countries. The organization's influence in the larger society — most significantly its directing elites — increases in turn.

The last line of defense against suicide terrorism — preventing bombers from reaching targets — may be the most expensive and least likely to succeed. Random bag or body searches cannot be very effective against people willing to die, although this may provide some semblance of security and hence psychological defense against suicide terrorism's psychological warfare. A middle line of defense, penetrating and destroying recruiting organizations and isolating their leaders, may be successful in the near term, but even more resistant organizations could emerge instead. The first line of defense is to drastically reduce receptivity of potential recruits to recruiting organizations. But how?

It is important to know what probably will not work. Raising literacy rates may have no effect and could be counterproductive should greater literacy translate into greater exposure to terrorist propaganda (in Pakistan, literacy and dislike for the United States increased as the number of religious madrasa schools increased from 3,000 to 39,000 since 1978). Lessening poverty may have no effect, and could be counterproductive if poverty reduction for the entire population amounted to a downward redistribution of wealth that left those initially better off with fewer opportunities than before. Ending occupation or reducing perceived humiliation may help, but not if the population believes this to be a victory inspired by terror (e.g., Israel's apparently forced withdrawal from Lebanon).

If suicide bombing is crucially (though not exclusively) an institution-level phenomenon, it may require finding the right mix of pressure and inducements to get the communities themselves to abandon support for institutions that recruit suicide attackers. One way is to so damage the community's social and political fabric that any support by the local population or authorities for sponsors of suicide attacks collapses, as happened regarding the kamikaze as a by-product of the nuclear destruction of Hiroshima and Nagasaki. In the present world, however, such a strategy would neither be morally justifiable nor practical to implement, given the dispersed and distributed organization of terrorist institutions among distantly separated populations that collectively number in the hundreds of millions. Likewise, retaliation in kind ("tit for tat") is not morally acceptable if allies are sought. Even in more localized settings, such as the Israeli-Palestinian conflict, coercive policies alone may not achieve lasting relief from attack and can exacerbate the problem over time. On the inducement side, social psychology research indicates that people who identify with antagonistic groups use conflicting information from the other group to reinforce antagonism. Thus, simply trying to persuade others from without by bombarding them with more self-serving information may only increase hostility.

Other research suggests that most people have more moderate views than what they consider their group norm to be. Inciting and empowering moderates from within to confront inadequacies and inconsistencies in their own knowledge (of others as evil), values (respect for life), and behavior (support for killing), and other members of their group, can produce emotional dissatisfaction leading to lasting change and influence on the part of these individuals. Funding for civic education and debate may help, also interfaith confidence building through intercommunity interaction initiatives (as Singapore's government proposes). Ethnic profiling, isolation, and preemptive attack on potential (but not yet actual) supporters of terrorism probably will not help. Another strategy is for the United States and its allies to change behavior by directly addressing and lessening sentiments of grievance and humiliation, especially in Palestine (where images of daily violence have made it the global focus of Moslem attention). For no evidence (historical or otherwise) indicates that support for suicide terrorism will evaporate without complicity in achieving at least some fundamental goals that suicide bombers and supporting communities share.

Of course, this does not mean negotiating over all goals, such as Al-Qaida's quest to replace the Western-inspired system of nation-states with a global caliphate, first in Moslem lands and then everywhere. Unlike other groups, Al-Qaida publicizes no specific demands after martyr actions. As with an avenging army, it seeks no compromise. But most people who currently sympathize with it might.

Perhaps to stop the bombing we need research to understand which configurations of psychological and cultural relationships are luring and binding thousands, possibly millions, of mostly ordinary people into the terrorist organization's martyr-making web. Study is needed on how terrorist institutions form and on similarities and differences across organizational structures, recruiting practices, and populations recruited. Are there reliable differences between religious and secular groups, or between ideologically driven and grievance-driven terrorism? Interviews with surviving Hamas bombers and captured Al-Qaida operatives suggest that ideology and grievance are factors for both groups but relative weights and consequences may differ.

We also need to investigate any significant causal relations between our society's policies and actions and those of terrorist organizations and supporters. We may find that the global economic, political, and cultural agenda of our own society has a catalyzing role in moves to retreat from our world view (Taliban) or to create a global counterweight (Al-Qaida). Funding such research may be difficult. As with the somewhat tendentious and self-serving use of

"terror" as a policy concept, to reduce dissonance our governments
and media may wish to ignore these relations as legitimate topics
for inquiry into what terrorism is all about and why it exists.

This call for research may demand more patience than any administration could politically tolerate during times of crisis. In the
long run, however, our society can ill afford to ignore either the
consequences of its own actions or the causes behind the actions of
others. Potential costs of such ignorance are terrible to contemplate. The comparatively minor expense of research into such consequences and causes could have inestimable benefit.

RONALD BAILEY

# The Battle for Your Brain

FROM *Reason*

"WE'RE ON THE VERGE of profound changes in our ability to ma-
nipulate the brain," says Paul Root Wolpe, a bioethicist at the University of Pennsylvania. He isn't kidding. The dawning age of neuroscience promises not just new treatments for Alzheimer's and
other brain diseases but enhancements to improve memory, boost
intellectual acumen, and fine-tune our emotional responses. "The
next two decades will be the golden age of neuroscience," declares
Jonathan Moreno, a bioethicist at the University of Virginia. "We're
on the threshold of the kind of rapid growth of information in neuroscience that was true of genetics fifteen years ago."

One man's golden age is another man's dystopia. One of the
more vociferous critics of such research is Francis Fukuyama, who
warns in his book *Our Posthuman Future* that "we are already in the
midst of this revolution" and "*we should use the power of the state to regulate it*" (emphasis his). In May a cover story in the usually pro-tech-
nology *Economist* worried that "neuroscientists may soon be able to
screen people's brains to assess their mental health, to distribute
that information, possibly accidentally, to employers or insurers,
and to 'fix' faulty personality traits with drugs or implants on demand."

There are good reasons to consider the ethics of tinkering directly with the organ from which all ethical reflection arises. Most
of those reasons boil down to the need to respect the rights of the
people who would use the new technologies. Some of the field's
moral issues are common to all biomedical research: how to design
clinical trials ethically, how to ensure subjects' privacy, and so on.

Others are peculiar to neurology. It's not clear, for example, whether people suffering from neurodegenerative disease can give informed consent to be experimented on.

Last May the Dana Foundation sponsored an entire conference at Stanford on "neuroethics." Conferees deliberated over issues like the moral questions raised by new brain scanning techniques, which some believe will lead to the creation of truly effective lie detectors. Participants noted that scanners might also be able to pinpoint brain abnormalities in those accused of breaking the law, thus changing our perceptions of guilt and innocence. Most nightmarishly, some worried that governments could one day use brain implants to monitor and perhaps even control citizens' behavior.

But most of the debate over neuroethics has not centered around patients' or citizens' autonomy, perhaps because so many of the field's critics themselves hope to restrict that autonomy in various ways. The issue that most vexes *them* is the possibility that neuroscience might enhance previously "normal" human brains.

The tidiest summation of their complaint comes from the conservative columnist William Safire. "Just as we have anti-depressants today to elevate mood," he wrote after the Dana conference, "tomorrow we can expect a kind of Botox for the brain to smooth out wrinkled temperaments, to turn shy people into extroverts, or to bestow a sense of humor on a born grouch. But what price will human nature pay for these nonhuman artifices?"

Truly effective neuropharmaceuticals that improve moods and sharpen mental focus are already widely available and taken by millions. While there is some controversy about the effectiveness of Prozac, Paxil, and Zoloft, nearly 30 million Americans have taken them, with mostly positive results. In his famous 1993 book *Listening to Prozac,* the psychiatrist Peter Kramer describes patients taking the drug as feeling "better than well." One Prozac user, called Tess, told him that when she isn't taking the medication, "I am not myself."

### One Pill Makes You Smarter...

That's exactly what worries Fukuyama, who thinks Prozac looks a lot like *Brave New World*'s soma. The pharmaceutical industry, he declares, is producing drugs that "provide self-esteem in the bottle by elevating serotonin in the brain." If you need a drug to be your "self," these critics ask, do you really have a self at all?

Another popular neuropharmaceutical is Ritalin, a drug widely prescribed to remedy attention deficit hyperactivity disorder (ADHD), which is characterized by agitated behavior and an inability to focus on tasks. Around 1.5 million schoolchildren take Ritalin, which recent research suggests boosts the activity of the neurotransmitter dopamine in the brain. Like all psychoactive drugs, it is not without controversy. Perennial psychiatric critic Peter Breggin argues that millions of children are being "drugged into more compliant or submissive state[s]" to satisfy the needs of harried parents and school officials. For Fukuyama, Ritalin is prescribed to control rambunctious children because "parents and teachers . . . do not want to spend the time and energy necessary to discipline, divert, entertain, or train difficult children the old-fashioned way."

Unlike the more radical Breggin, Fukuyama acknowledges that drugs such as Prozac and Ritalin have helped millions when other treatments have failed. Still, he worries about their larger social consequences. "There is a disconcerting symmetry between Prozac and Ritalin," he writes. "The former is prescribed heavily for depressed women lacking in self-esteem; it gives them more the alpha-male feeling that comes with high serotonin levels. Ritalin, on the other hand, is prescribed largely for young boys who do not want to sit still in class because nature never designed them to behave that way. Together, the two sexes are gently nudged toward that androgynous median personality, self-satisfied and socially compliant, that is the current politically correct outcome in American society."

Although there are legitimate questions here, they're related not to the chemicals themselves but to who makes the decision to use them. Even if Prozac and Ritalin can help millions of people, that doesn't mean schools should be able to force them on any student who is unruly or bored. But by the same token, even if you accept the most radical critique of the drug — that ADHD is not a real disorder to begin with — that doesn't mean Americans who exhibit the symptoms that add up to an ADHD diagnosis should not be allowed to alter their mental state chemically, if that's an outcome they want and a path to it they're willing to take.

Consider Nick Megibow, a senior majoring in philosophy at Gettysburg College. "Ritalin made my life a lot better," he reports. "Before I started taking Ritalin as a high school freshman, I was doing really badly in my classes. I had really bad grades, Cs and Ds mostly. By sophomore year, I started taking Ritalin, and it really worked amazingly. My grades improved dramatically to mostly As and Bs. It allows me to focus and get things done rather than take three times the amount of time that it should take to finish something." If people like Megibow don't share Fukuyama's concerns about the wider social consequences of their medication, it's because they're more interested, quite reasonably, in feeling better and living a successful life.

What really worries critics like Safire and Fukuyama is that Prozac and Ritalin may be the neuropharmacological equivalent of bearskins and stone axes compared to the new drugs that are coming. Probably the most critical mental function to be enhanced is memory. And this, it turns out, is where the most promising work is being done. At Princeton, biologist Joe Tsien's laboratory famously created smart mice by genetically modifying them to produce more NMDA brain receptors, which are critical for the formation and maintenance of memories. Tsien's mice were much faster learners than their unmodified counterparts. "By enhancing learning, that is, memory acquisition, animals seem to be able to solve problems faster," notes Tsien. He believes his work has identified an important target that will lead other researchers to develop drugs that enhance memory.

A number of companies are already hard at work developing memory drugs. Cortex Pharmaceuticals has developed a class of compounds called AMPA receptor modulators, which enhance the glutamate-based transmission between brain cells. Preliminary results indicate that the compounds do enhance memory and cognition in human beings. Memory Pharmaceuticals, cofounded by Nobel laureate Eric Kandel, is developing a calcium channel receptor modulator that increases the sensitivity of neurons and allows them to transmit information more speedily and a nicotine receptor modulator that plays a role in synaptic plasticity. Both modulators apparently improve memory. Another company, Targacept, is working on the nicotinic receptors as well.

All these companies hope to cure the memory deficits that some

30 million baby boomers will suffer as they age. If these compounds can fix deficient memories, it is likely that they can enhance normal memories as well. Tsien points out that a century ago the encroaching senility of Alzheimer's disease might have been considered part of the "normal" progression of aging. "So it depends on how you define *normal*," he says. "Today we know that most people have less good memories after age forty, and I don't believe that's a normal process."

## Eight Objections

And so we face the prospect of pills to improve our mood, our memory, our intelligence, and perhaps more. Why would anyone object to that?

Eight objections to such enhancements recur in neuroethicists' arguments. None of them is really convincing.

*Neurological enhancements permanently change the brain.* Erik Parens of the Hastings Center, a bioethics think tank, argues that it's better to enhance a child's performance by changing his environment than by changing his brain — that it's better to, say, reduce his class size than to give him Ritalin. But this is a false dichotomy. Reducing class size is aimed at changing the child's biology too, albeit indirectly. Activities like teaching are supposed to induce biological changes in a child's brain, through a process called *learning*.

Fukuyama falls into this same error when he suggests that even if there is some biological basis for their condition, people with ADHD "clearly . . . can do things that would affect their final degree of attentiveness or hyperactivity. Training, character, determination, and environment more generally would all play important roles." So can Ritalin, and much more expeditiously, too. "What is the difference between Ritalin and the Kaplan SAT review?" asks the Dartmouth neuroscientist Michael Gazzaniga. "It's six of one and a half dozen of the other. If both can boost SAT scores by, say, 120 points, I think it's immaterial which way it's done."

*Neurological enhancements are antiegalitarian.* A perennial objection to new medical technologies is the one Parens calls "unfairness in the distribution of resources." In other words, the rich and their children will get access to brain enhancements first, and will thus acquire more competitive advantages over the poor.

This objection rests on the same false dichotomy as the first. As the University of Virginia's Moreno puts it, "We don't stop people from giving their kids tennis lessons." If anything, the new enhancements might *increase* social equality. Moreno notes that neuropharmaceuticals are likely to be more equitably distributed than genetic enhancements, because "after all, a pill is easier to deliver than DNA."

*Neurological enhancements are self-defeating.* Not content to argue that the distribution of brain enhancements won't be egalitarian enough, some critics turn around and argue that it will be *too* egalitarian. Parens has summarized this objection succinctly: "If everyone achieved the same relative advantage with a given enhancement, then ultimately no one's position would change; the 'enhancement' would have failed if its purpose was to increase competitive advantage."

This is a flagrant example of the zero-sum approach that afflicts so much bioethical thought. Let's assume, for the sake of argument, that everyone in society will take a beneficial brain-enhancing drug. Their relative positions may not change, but the overall productivity and wealth of society would increase considerably, making everyone better off. Surely that is a social good.

*Neurological enhancements are difficult to refuse.* Why exactly would everyone in the country take the same drug? Because, the argument goes, competitive pressures in our go-go society will be so strong that a person will be forced to take a memory-enhancing drug just to keep up with everyone else. Even if the law protects freedom of choice, social pressures will draw us in.

For one thing, this misunderstands the nature of the technology. It's not simply a matter of popping a pill and suddenly zooming ahead. "I know a lot of smart people who don't amount to a row of beans," says Gazzaniga. "They're just happy underachieving, living life below their potential. So a pill that pumps up your intellectual processing power won't necessarily give you the drive and ambition to use it."

Beyond that, it's not as though we don't all face competitive pressures anyway — to get into and graduate from good universities, to constantly upgrade skills, to buy better computers and more productive software, whatever. Some people choose to enhance themselves by getting a Ph.D. in English; others are happy to stop their formal education after high school. It's not clear why a pill should be more irresistible than higher education, or why one should raise special ethical concerns while the other does not.

*Neurological enhancements undermine good character.* For some critics, the comparison to higher education suggests a different problem. We should strive for what we get, they suggest; taking a pill to enhance cognitive functioning is just too easy. As Fukuyama puts it: "The normal, and morally acceptable, way of overcoming low self-esteem was to struggle with oneself and with others, to work hard, to endure painful sacrifices, and finally to rise and be seen as having done so."

"By denying access to brain-enhancing drugs, people like Fukuyama are advocating an exaggerated stoicism," counters Moreno. "I don't see the benefit or advantage of that kind of tough love." Especially since there will still be many different ways to achieve things and many difficult challenges in life. Brain-enhancing drugs might ease some of our labors, but as Moreno notes, "there are still lots of hills to climb, and they are pretty steep." Cars, computers, and washing machines have tremendously enhanced our ability to deal with formerly formidable tasks. That doesn't mean life's struggles have disappeared — just that we can now tackle the next ones.

*Neurological enhancements undermine personal responsibility.* Carol Freedman, a philosopher at Williams College, argues that what is at stake "is a conception of ourselves as responsible agents, not machines." Fukuyama extends the point, claiming that "ordinary people" are eager to "medicalize as much of their behavior as possible and thereby reduce their responsibility for their own actions." As an example, he suggests that people who claim to suffer from ADHD "want to absolve themselves of personal responsibility."

But we are not debating people who might use an ADHD diagnosis as an excuse to behave irresponsibly. We are speaking of people who use Ritalin to *change* their behavior. Wouldn't it be more irresponsible of them to not take corrective action?

*

*Neurological enhancements enforce dubious norms.* There are those who assert that corrective action might be irresponsible after all, depending on just what it is that you're trying to correct. People might take neuropharmaceuticals, some warn, to conform to a harmful social conception of normality. Many bioethicists — Georgetown University's Margaret Little, for example — argue that we can already see this process in action among women who resort to expensive and painful cosmetic surgery to conform to a social ideal of feminine beauty. Never mind for the moment that beauty norms for both men and women have never been so diverse. Providing and choosing to avail oneself of that surgery makes one complicit in norms that are morally wrong, the critics argue. After all, people should be judged not by their physical appearances but by the content of their characters.

That may be so, but why should someone suffer from society's slights if she can overcome them with a nip here and a tuck there? The norms may indeed be suspect, but the suffering is experienced by real people whose lives are consequently diminished. Little acknowledges this point, but argues that those who benefit from using a technology to conform have a moral obligation to fight against the suspect norm. Does this mean people should be given access to technologies they regard as beneficial only if they agree to sign on to a bioethical fatwa?

Of course, we should admire people who challenge norms they disagree with and live as they wish, but why should others be denied relief just because some bioethical commissars decree that society's misdirected values must change? Change may come, but real people should not be sacrificed to some restrictive bioethical utopia in the meantime. Similarly, we should no doubt value depressed people or people with bad memories just as highly as we do happy geniuses, but until that glad day comes people should be allowed to take advantage of technologies that improve their lives in the society in which they actually live.

Furthermore, it's far from clear that everyone will use these enhancements in the same ways. There are people who alter their bodies via cosmetic surgery to bring them closer to the norm, and there are people who alter their bodies via piercings and tattoos to make them more individually expressive. It doesn't take much imagination to think of unusual or unexpected ways that Americans might use mind-enhancing technologies. Indeed, the war on drugs is being waged, in part, against a small but significant minority of people who prefer to alter their consciousness in socially disapproved ways.

*Neurological enhancements make us inauthentic.* Parens and others worry that the users of brain-altering chemicals are less authentically themselves when they're on the drug. Some of them would reply that the exact opposite is the case. In *Listening to Prozac,* Kramer chronicles some dramatic transformations in the personalities and attitudes of his patients once they're on the drug. The aforementioned Tess tells him it was "as if I had been in a drugged state all those years and now I'm clearheaded."

Again, the question takes a different shape when one considers the false dichotomy between biological and "nonbiological" enhancements. Consider a person who undergoes a religious conversion and emerges from the experience with a more upbeat and attractive personality. Is he no longer his "real" self? Must every religious convert be deprogrammed?

Even if there were such a thing as a "real" personality, why should you stick with it if you don't like it? If you're socially withdrawn and a pill can give you a more vivacious and outgoing manner, why not go with it? After all, you're choosing to take responsibility for being the "new" person the drug helps you to be.

## Authenticity and Responsibility

"Is it a drug-induced personality or has the drug cleared away barriers to the real personality?" asks the University of Pennsylvania's Wolpe. Surely the person who is choosing to use the drug is in a better position to answer that question than some bioethical busybody.

This argument over authenticity lies at the heart of the neuroethicists' objections. If there is a single line that divides the supporters of neurological freedom from those who would restrict the new treatments, it is the debate over whether a natural state of human being exists and, if so, how appropriate it is to modify it. Wolpe makes the point that in one sense cognitive enhancement resembles its opposite, Alzheimer's disease. A person with Alzheimer's loses her personality. Similarly, an enhanced individual's personality may become unrecognizable to those who knew her before.

Not that this is unusual. Many people experience a version of this process when they go away from their homes to college or the military. They return as changed people with new capacities, likes, dislikes, and social styles, and they often find that their families and friends no longer relate to them in the old ways. Their brains have been changed by those experiences, and they are not the same people they were before they went away. Change makes most people uncomfortable, probably never more so than when it happens to a loved one. Much of the neuro-Luddites' case rests on a belief in an unvarying, static personality, something that simply doesn't exist.

It isn't just personality that changes over time. Consciousness itself is far less static than we've previously assumed, a fact that raises contentious questions of free will and determinism. Neuroscientists are finding more and more of the underlying automatic processes operating in the brain, allowing us to take a sometimes disturbing look under our own hoods. "We're finding out that by the time we're conscious of doing something, the brain's already done it," explains Gazzaniga. Consciousness, rather than being the director of our activities, seems instead to be a way for the brain to explain to itself why it did something.

Haunting the whole debate over neuroscientific research and neuroenhancements is the fear that neuroscience will undercut notions of responsibility and free will. Very preliminary research has suggested that many violent criminals do have altered brains. At the Stanford conference, *Science* editor Donald Kennedy suggested that once we know more about brains, our legal system will have to make adjustments in how we punish those who break the law. A murderer or rapist might one day plead innocence on the grounds that "my amygdala made me do it." There is precedent for this: The legal system already mitigates criminal punishment when an offender can convince a jury he's so mentally ill that he cannot distinguish right from wrong.

Of course, there are other ways such discoveries might pan out in the legal system, with results less damaging to social order but still troubling for notions of personal autonomy. One possibility is that an offender's punishment might be reduced if he agrees to take a pill that corrects the brain defect he blames for his crime. We already hold people responsible when their drug use causes harm to others — most notably, with laws against drunk driving. Perhaps in the future we will hold people responsible if they fail to take drugs that would help prevent them from behaving in harmful ways. After all, which is more damaging to personal autonomy, a life confined to a jail cell or roaming free while taking a medication?

The philosopher Patricia Churchland examines these conundrums in her forthcoming book, *Brainwise: Studies in Neurophilosophy.* "Much of human social life depends on the expectation that agents have control over their actions and are responsible for their choices," she writes. "In daily life it is commonly assumed that it is sensible to punish and reward behavior so long as the person was in control and chose knowingly and intentionally." And that's the way it should remain, even as we learn more about how our brains work and how they sometimes break down.

Churchland points out that neuroscientific research by scientists like the University of Iowa's Antonio Damasio strongly shows that emotions are an essential component of viable practical reasoning about what a person should do. In other words, neuroscience is bolstering philosopher David Hume's insight that "reason is and ought only to be the slave of the passions." Patients whose affects are depressed or lacking due to brain injury are incapable of judging or evaluating between courses of action. Emotion is what prompts and guides our choices.

Churchland further argues that moral agents come to be morally and practically wise not through pure cognition but by developing moral beliefs and habits through life experiences. Our moral reflexes are honed through watching and hearing about which actions are rewarded and which are punished; we learn to be moral the same way we learn language. Consequently, Churchland concludes "the default presumption that agents are responsible for their actions is empirically necessary to an agent's learning, both emotionally and cognitively, how to evaluate the consequences of certain events and the price of taking risks."

It's always risky to try to derive an "ought" from an "is," but neuroscience seems to be implying that liberty — i.e., letting people make choices and then suffer or enjoy the consequences — is essential for inculcating virtue and maintaining social cooperation. Far from undermining personal responsibility, neuroscience may end up strengthening it.

## For Neurological Liberty

Fukuyama wants to "draw red lines" to distinguish between therapy and enhancement, "directing research toward the former while putting restrictions on the latter." He adds that "the original purpose of medicine is, after all, to heal the sick, not turn healthy people into gods." He imagines a federal agency that would oversee neurological research, prohibiting anything that aims at enhancing our capacities beyond some notion of the human norm.

"For us to flourish as human beings, we have to live according to our nature, satisfying the deepest longings that we as natural beings have," Fukuyama told the Christian review *Books & Culture* last summer. "For example, our nature gives us tremendous cognitive capabilities, capability for reason, capability to learn, to teach ourselves things, to change our opinions, and so forth. What follows from that? A way of life that permits such growth is better than a life in which this capacity is shriveled and stunted in various ways." This is absolutely correct. The trouble is that Fukuyama has a shriveled, stunted vision of human nature, leading him and others to stand athwart neuroscientific advances that will make it possible for more people to take fuller advantage of their reasoning and learning capabilities.

Like any technology, neurological enhancements can be abused, especially if they're doled out — or imposed — by an unchecked authority. But Fukuyama and other critics have not made a strong case for why *individuals,* in consultation with their doctors, should not be allowed to take advantage of new neuroscientific breakthroughs to enhance the functioning of their brains. And it is those individuals that the critics will have to convince if they seriously expect to restrict this research.

It's difficult to believe that they'll manage that. In the 1960s many states outlawed the birth control pill, on the grounds that it would be too disruptive to society. Yet Americans, eager to take control of their reproductive lives, managed to roll back those laws, and no one believes that the pill could be re-outlawed today.

Moreno thinks the same will be true of the neurological advances to come. "My hunch," he says, "is that in the United States, medications that enhance our performance are not going to be prohibited." When you consider the sometimes despairing tone that Fukuyama and others like him adopt, it's hard not to conclude that on that much, at least, they agree.

PHILIP M. BOFFEY

# Fearing the Worst Should Anyone Produce a Cloned Baby

FROM *The New York Times*

EXPERTS INCREASINGLY suspect that a fringe cult's claim to have cloned a human baby is a publicity-seeking hoax, especially now that the group seems to be evading genetic tests that might prove its claim. The cult's deep-seated belief that space aliens created the human race by cloning is so wacko that all of its other claims become suspect. But with several renegade groups supposedly racing to produce the first cloned baby, it is almost inevitable that sooner or later someone will succeed. It's time to start preparing ourselves mentally for that eventuality.

Until now, there has been widespread support in Congress for a ban on cloning to produce babies — the only real debate has been over the use of cloned embryos to find cures for disease. Even scientists who want to pursue therapeutic cloning have been happy to endorse a ban on reproductive cloning for safety reasons. Judging from animal tests, there is just too much risk that a cloned baby would be born with birth defects or face medical problems as it ages. But if the renegade cloners ever present a healthy baby who is shown by genetic tests to be a cloned copy of an adult, the safety argument will become less persuasive. It will then be imperative to look much harder at the ethical and moral implications of reproductive cloning.

The public's fear and fascination with cloning, as expressed in popular culture, focuses on some highly improbable scenarios. *The Boys from Brazil*, a 1978 movie based on a novel, featured a plot by

Nazi doctors to produce a cadre of young Hitler clones to start a Fourth Reich. Although the villains tried to give the young Hitlers the same home life as the original, the writers slid past the overwhelming probability that even Hitler himself, introduced into a different historical context, would not have the same career trajectory.

The most recent *Star Wars* movie, last year's *Attack of the Clones,* featured an army of clones derived from the genes of an aggressive bounty hunter, modified to ensure willingness to follow military orders. The image of a horde of unthinking, cloned attackers is a classic science-fiction nightmare. But producing such an army with today's techniques would require a huge number of women to supply the eggs and bear the fetal clones to term, a problem that is often glossed over in horror stories. On a more individual scale, the new *Star Trek* movie, *Nemesis,* pits a good spaceship captain against his evil younger clone. The younger version went bad because of his harsh treatment when exiled to hellish mines on a remote world, a nice reminder that genes alone do not dictate destiny.

Over the years, people have fretted that cloning practiced widely might eliminate the need for men (women could bear children asexually), might exacerbate the male-female ratio (cultures that revere males could clone only them), and might reduce the genetic diversity that comes from mingling genes in sexual reproduction. But those scenarios suppose that cloning might indeed become the preferred means of reproduction, an event that would seem to require an unlikely mass change in human preferences or a totalitarian regime to impose its will. In a democratic, free-market society, commercial entities might well promote cloning, but such marketing has not turned in vitro fertilization or the freezing of bodies for later resurrection into mass commodities.

Some critics fear that cloning could usher in a new eugenics, in which nations or individuals might try to improve the average capabilities of the next generation by cloning the likes of an Einstein, Mozart, Michael Jordan, or Marilyn Monroe. Such genetic enhancement has in fact been endorsed by some eminent scientists in the past, and no doubt there are individuals who might want offspring with particular talents. Yet sperm banks with seed from famous and accomplished men have existed for some time with no sign of a mass rush to use them.

In its report on human cloning last year, the President's Council on Bioethics worried that cloning to produce children could disrupt the normal relationships between generations and within families, could turn children into manufactured products rather than independent beings, and could put undue pressure on a cloned child living in the shadow of a genetically identical adult. Most of the panel's concerns were necessarily speculative, and some of its worries seem overdrawn. Twins seem to do just fine with the same genome, for example, so it is not clear that having a twin a generation older would be all that burdensome. Nor is it clear that families with a cloned child would face more confusing relationships than already exist in today's divorced, blended, and extended families. On the other hand, cloning could prove medically useful for couples worried about passing on genetic diseases, infertile individuals who could not have a biologically related child any other way, or parents needing a compatible tissue donor to cure a sick child.

For the immediate future, Congress would be wise to ban reproductive cloning as far too risky while allowing therapeutic cloning to proceed. But sooner or later technical advances may diminish the risks. The nation needs to focus on what to do then.

AUSTIN BUNN

# The Bittersweet Science

FROM *The New York Times Magazine*

ELEVEN-YEAR-OLD Elizabeth Hughes was, in retrospect, the ideal patient: bright, obedient, uncomplaining, and wholly unprepared to die. Born in 1907 in the New York State governor's mansion, Elizabeth was the daughter of Charles Evans Hughes, who later became a justice on the Supreme Court, ran against Woodrow Wilson in 1916, and served as secretary of state under Harding. Elizabeth had a perfectly normal, aristocratic youth until she seemed to become allergic to childhood. She would come home from friends' birthday parties with an insatiable thirst, drinking almost two quarts of water at a sitting. By winter, she had become thin, constantly hungry, and exhausted. Her body turned into a sieve: No matter how much water she drank, she was always thirsty.

In early 1919, Elizabeth's parents took her to a mansion in Morristown, New Jersey, recently christened the Physiatric Institute and run by Dr. Frederick Allen. A severe, debt-ridden clinician with a pockmarked résumé, Allen had written the authoritative account on treating her condition. He prolonged hundreds of lives and was the girl's best chance. Allen examined Elizabeth and diagnosed diabetes — her body was not properly processing her food into fuel — and told her parents what they would never tell their daughter: that her life expectancy was one year, three at the outside. Even that was a magnificent extension of previous fatality rates. "The diagnosis was like knowing a death sentence had been passed," wrote one historian. Then Dr. Allen did what many doctors at the time would have done for Elizabeth, except that this doctor was exceptionally good, if not the finest in the world, at it. He began to starve her.

The history of medicine "is like the night sky," says the historian Roy Porter in his book *The Greatest Benefit to Mankind: A Medical History of Humanity.* "We see a few stars and group them into mythic constellations. What is chiefly visible is the darkness."

Diabetes doesn't come from simply eating too much sugar; nor is it cured, as was once thought, by a little horseback riding. It is not the result of a failing kidney, overactive liver, or phlegmy disposition, though these were the authoritative answers for centuries. Diabetes happens when the blood becomes saturated with glucose, the body's main energy source, which is normally absorbed by the cells — which is to say that the pathology of diabetes is subtle and invisible, so much so that a third of the people who have it don't even know it. Until the prohibition against autopsies was gradually lifted (by 1482, the pope had informally sanctioned it), what we knew of human anatomy came through the tiny window of war wounds and calamitous gashes — and even then it took centuries for doctors to decide just what the long, lumpy organ called the pancreas actually did or, in the case of diabetes, didn't do. We like to think surgically about the history of medicine, that it moved purposefully from insight to insight, angling closer to cure. But that is only the luxury of contemporary life. Looked at over time, medicine doesn't advance as much as grope forward, with remedies — like bloodletting; quicksilver ointments; and simple, unendurable hunger — that blurred the line between treatment and torture.

Diabetes was first diagnosed by the Greek physician Aretaeus of Cappadocia, who deemed it a "wonderful affection . . . being a melting down of the flesh and limbs into urine." For the afflicted, "life is disgusting and painful; thirst unquenchable . . . and one cannot stop them from drinking or making water." Since the classical period forbade dissection, Porter notes, "hidden workings had to be deduced largely from what went in and what came out." An early diagnostic test was to swill urine, and to the name *diabetes,* meaning "siphon," was eventually added *mellitus,* meaning "sweetened with honey." Healers could often diagnose diabetes without the taste test. Black ants were attracted to the urine of those wasting away, drawn by the sugar content. Generations later, doctors would make a similar deduction by spotting dried white sugar spots on the shoes or pants of diabetic men with bad aim.

For the Greeks, to separate disease symptoms from individual pain while isolating them from magical causes was itself an enormous intellectual leap. "We should be really impressed with Aretaeus," says Dr. Chris Feudtner, author of the coming *Bittersweet: Diabetes, Insulin, and the Transformation of Illness.* "He was able to spot the pattern of diabetes in a dense thicket of illness and suffering."

But for centuries, this increasing precision in disease recognition was not followed by any effective treatment — more details didn't make physicians any less helpless. At the time, they were unknowingly confusing two kinds of diabetes: Type 1, known until recently as "juvenile diabetes," which is more extreme but less common than Type 2, or "adult onset," which seems to be related to obesity and overeating. With Type 1 (what Elizabeth Hughes had), the pancreas stops secreting insulin, a hormone that instructs the body to use the sugar in the blood for energy. With Type 2, the pancreas produces insulin (at least initially), but the tissues of the body stop responding appropriately. By 1776, doctors were still just boiling the urine of diabetics to conclusively determine that they were passing sugar, only to watch their patients fall into hyperglycemic comas and die.

If dangerous levels of glucose were pumping out of diabetics, one idea was obvious: Stop it from going in. That demanded a more sophisticated understanding of food itself. In the long tradition of grotesque scientific experimentation, an insight came through a lucky break: a gaping stomach wound. In 1822, William Beaumont, a surgeon in the U.S. Army, went to the Canadian border to treat a nineteen-year-old trapper hit by a shotgun. The boy recovered, but he was left with a hole in his abdomen. According to Porter, Beaumont "took advantage of his patient's unique window" and dropped food in on a string. The seasoned beef took the longest to digest. Stale bread broke down the quickest. The digestion process clearly worked differently depending on what was eaten. Then during the 1871 siege of Paris by the Germans, a French doctor named Apollinaire Bouchardat noticed that, though hundreds were starving to death, his diabetic patients strangely improved. This became the basis for a new standard of treatment. *Mangez le moins possible,* he advised them. Eat as little as possible.

In the spring of 1919, when Elizabeth Hughes came under Dr. Allen's care, she weighed seventy-five pounds and was nearly five feet tall. For one week, he fasted her. Then he put her on an extremely low-calorie diet to eradicate sugar from her urine. If the normal caloric intake for a girl her age is between 2,200 and 2,400 calories daily, Elizabeth took in 400 to 600 calories a day for several weeks, including one day of fasting each week. Her weight, not surprisingly, plummeted. As Michael Bliss notes in his book *The Discovery of Insulin*, the Hughes family brought in a nurse to help weigh and supervise every gram of food that she ate. Desserts and bread were verboten. "She lived on lean meat, eggs, lettuce, milk, a few fruits, tasteless bran rusks, and tasteless vegetables (boiled three times to make them almost totally carbohydrate-free)," Bliss writes. Instead of a birthday cake, she had to settle for "a hat box covered in pink and white paper with candles on it. On picnics in the summertime she had her own little frying pan to cook her omelet in while the others had chops, fresh fish, corn on the cob, and watermelon."

You could say that Elizabeth Hughes was on a twisted precursor of the Zone diet: Her menu relied on proteins and fats, with the abolishment of carbohydrates like bread and pasta. In fact, Allen's maniacal scrutiny of his patients' nutrition — fasting them, weighing each meal, counting calories — was one of the first "diets" in the modern sense. At the time Elizabeth entered the clinic, being well fed was a sign of good health. But the new science of nutrition fostered the idea of weight reduction as a standard of health and not illness.

Allen's "starvation diet" was a particular cruelty. Patients came to him complaining of hunger and rapid weight loss, and Allen demanded further restrictions, further weight loss. "Yes, the method was severe; yes, many patients could not or would not follow it," writes Bliss. "But what was the alternative?" Over the years, doctors recommended opium, even heaps of sugar (which only accelerated death, but since nothing else worked, why not enjoy the moment?). But nobody had a better way than Allen to extend lives. If the fasting wasn't working and symptoms got worse, Allen insisted on more rigorous undernourishment. In his campaigns to master their disease, Allen took his patients right to the edge of death, but he justified this by pointing out that patients faced a stark choice: die of diabetes or risk "inanition," which Allen explained as "starvation due to inability to acquire tolerance for any living diet." The Physiatric Institute became a famine ward.

Some of Allen's patients survived levels of inanition not thought possible, Bliss writes. One twelve-year-old patient, blind from diabetes when he was admitted, still occasionally showed sugar in his urine. The clinic became convinced that the kid — so weak he could barely get out of bed — was somehow stealing food. "It turned out that his supposed helplessness was the very thing that gave him opportunities which other persons lacked," Allen later wrote in his book, *Total Dietary Regulation in the Treatment of Diabetes.* "Among unusual things eaten were toothpaste and birdseed, the latter being obtained from the cage of a canary which he had asked for." The staff, thinking he was pilfering food, cut his diet back and further back. The boy weighed less than forty pounds when he died from starvation.

No one explained to Elizabeth Hughes why the friends she made at Allen's clinic stopped writing her letters. Death was kept hidden, though it must have been obvious from the halls of the clinic, where rows of gaunt children stared from their beds. "It would have been unendurable if only there had not been so many others," one Allen nurse wrote. Dutifully, Elizabeth — strong enough just to read and sew — hardly ever showed sugar. Her attendant punished her severely the one time she caught her stealing turkey skin from the kitchen after Thanksgiving. Still, she was wasting away. By April 1921, thirteen years old and two years into her treatment, Elizabeth was down to fifty-two pounds and averaged 405 calories a day. In letters to her parents, she talked about getting married and what she would do on her twenty-first birthday. Reading the letters "must have been heartbreaking," writes Bliss. "Elizabeth was a semi-invalid."

In the history of illness, there are countless medicines, over time and across cultures, with varying degrees of suffering and success. There is only one kind of cure — the one that invariably, irrefutably works. Insulin is not a cure. It is a treatment, but it changed everything. In the summer of 1922, two young clinicians in Toronto named Frederick Banting and Charles Best surgically removed the pancreases from dozens of dogs, causing the dogs to "get" diabetes. They found that by injecting the dogs with a filtered solution of macerated pancreas (either the dogs' own or from calf fetuses), the glucose level in the dogs' blood dropped to normal. The researchers had discovered insulin.

But in August 1922, Dr. Frederick Allen had patients who could not wait, like Elizabeth. Allen left for Toronto to secure insulin. While he was gone, word leaked through his clinic about the breakthrough. Patients "who had not been out of bed for weeks began to trail weakly about, clinging to walls and furniture," wrote one nurse. "Big stomachs, skin-and-bone necks, skull-like faces . . . they looked like an old Flemish painter's depiction of a resurrection after famine. It was a resurrection, a crawling stirring, as of some vague springtime."

On the night Allen returned to the clinic, he found his patients — "silent as the bloated ghosts they looked like" — waiting in the hallway for him, wrote the nurse. "When he appeared through the open doorway, he caught the full beseeching of a hundred pair of eyes. It stopped him dead. Even now I am sure it was minutes before he spoke to them. . . . 'I think,' he said. 'I think we have something for you.'" He did, but not nearly enough. Though the results were striking — with the insulin, sugar vanished from the urine of "some of the most hopelessly severe cases of diabetes I have ever seen," wrote Allen — he did not have enough extract to treat all his patients, including Elizabeth. So her parents got her to Toronto. When Banting saw Elizabeth, she was three days away from her fifteenth birthday. She weighed forty-five pounds. He wrote: "Patient extremely emaciated . . . hair brittle and thin . . . muscles extremely wasted. . . . She was scarcely able to walk."

He started her insulin treatment immediately. The first injections cleared the sugar from her urine, and by the end of the first week, she was up to 1,220 calories a day, still without sugar. By the next, she was at 2,200 calories. Banting advised her to eat bread and potatoes, but she was incredulous. It had been three and a half years since she had them. That fall, she was one of several hundred North American diabetics pulled back from the edge. By November, she went home to her parents in Washington, and by January, she weighed 105 pounds. The same year, the thirty-one-year-old Banting won the Nobel Prize. Meanwhile, Dr. Allen, proprietor of an expensive clinic whose patients no longer needed him, went broke. Insulin was a miracle drug, resurrecting diabetics from comas and putting flesh on skeletons and, since it needed to be administered at least twice daily, it was a miracle that would be performed over and over. The era of chronic medical care had begun.

That may be the most poignant part of the history of Allen's clinic. The end of the famine of Elizabeth Hughes is really the start of another hunger: for the drugs that will keep us well for the rest of our lives. Elizabeth went to Barnard, reared three children, drank and smoked, but kept her diabetes a secret almost her entire life. She died of a heart attack in 1981, more than 43,000 injections of insulin later. But if the discovery of insulin took away the terror of diabetes, it replaced the miraculous with the routine. Healing lost one major ingredient: awe. "To think that I'll be leading a normal, healthy existence is beyond all comprehension," Elizabeth wrote to her mother, days after her first injection, in 1922. "It is simply too wonderful for words."

JENNET CONANT

# The New Celebrity

FROM *Seed*

JAMES DEWEY WATSON prefers to lecture no more than once a month (with a few exceptions), frequently testifies before Congress on behalf of genetic testing (but says he never tells them what they want to hear), and occasionally grants interviews like this one, but only in his office at Cold Spring Harbor Laboratory, and only if he can talk about what he wants to talk about on that given day. He does not have an e-mail address, which he says "cuts out 95 percent of the chatter," and unlike many of his famous peers, only sits on a handful of corporate boards. Not because he is against scientists sitting on boards in principle, but because Johnson & Johnson and Pfizer have never asked him to, "as I suspect I am known as 'not a good committee person.'"

Wherever Watson goes he is asked momentous questions about the future of humanity, fetal cloning, and genetically designed babies, and while he is willing to entertain most inquiries seriously, he is unwilling to take himself too seriously in the process. "Most institutions like to have Nobel Prize–winners around because it implies a certain standard," he says, straightening up in his chair and comically thrusting his chin forward with an air of utter pomposity. "But it's overrated in most cases. The local Nobel Prize–winner is usually pretty stale, as he probably did most of the work forty years before."

It's easy to see why Watson's sharp tongue has not always endeared him to his colleagues, and he cheerfully admits many people found him "quite unbearable" when he was young. Never able to resist the impulse to show off, he has forged a career that is equal parts brilliant intellectual opportunist and "outrageous prankster," as one biographer put it. Both in person and on the page, he is engaging, irreverent, and extraordinarily — at times, exasperatingly — provocative, as though determined to keep the spotlight intact.

At seventy-four, he is wildly popular with audiences, commanding $25,000 on the celebrity speaking circuit. He has a publicist and shares a big-name lecture agent with Bill Clinton. He spends much of his time raising money for his Cold Spring Harbor Laboratory, though confesses to embarking on a series of well-paid appearances to pay for the $100,000 "Twisting Dendrites" chandelier designed by Dale Chihuly that now hangs in one of the neuroscience labs. "It was too good a deal to pass up," he says sheepishly. He also footed the bill for Ballybung, the strikingly dramatic, orange-hued, Palladio-style president's house he had designed for him by the British architect Bill Grover, and which he plans to leave to the laboratory.

Watson has been a celebrated scientist for so long now that when he gives lectures, he spends most of his time expounding on his personal theory on how to achieve scientific success, otherwise known as, "Why I Deserved to Discover the Structure of DNA." He's boiled his formula for career advancement down to five easy steps, or "criteria," as he calls them: Number One: "Go for broke." This is why he picked DNA. "If you are going to do important science, do it." Number Two: "Have a way to get the answer." If you haven't a clue, you're going to waste time. Number Three: "Be obsessive." He not only knew DNA was important, it was all he could think about night and day. "Did you see Jeff Goldblum play me in the BBC film? Crick didn't get cast right; he didn't come across in any way as obsessive, whereas I did. It was DNA or nothing for me." Number Four: "Be part of a team." Working with Crick, he had a partner to bounce ideas off and a pal to support him. Number Five: "Talk to your opponents." A lot of scientists are afraid to share their ideas. But by cooperating with Maurice Wilkins, a scientist at a rival lab in London, Watson and Crick learned of experimental evidence that enabled them to clinch their discovery. The person who actually took the pioneering photograph, Rosalind Franklin, never shared her research, and died four years before the 1962 Nobel Prize was awarded to the three men. "Generally, it pays to talk," says Watson. "Oh, and another rule: Never be the brightest person in the room; then you can't learn anything."

Abruptly, class is dismissed. Watson is on to the next thing. That's how his mind works, darting from idea to idea, all in revolution around his interests and feelings and impulses. I play catch-up, following his glance, trying to figure out where he is going. "The photo shoot yesterday was quite interesting because I don't think I've ever posed for that kind of picture before," Watson reflects, in a voice so low that it's barely audible, making it necessary for me to nudge the tape recorder across the polished oak desk, past the spread of medals and prizes lined up along the window sill, the Nobel among them.

"That one was taken by *Vogue* magazine in June of 1954," he continues, his head cocked at a forty-five-degree angle as he contemplates a large, arty, black-and-white portrait of the Scientist as a Young Man that is propped on the floor against the far wall. "It was for an issue devoted to 'hot young people,' so they had Richard Burton and me. I was twenty-six."

Spinning around in his swivel chair at an inadvisable speed, Watson brushes wispy gray hair out of his pale blue eyes to survey the trophy wall behind his desk. It's cluttered with all manner of celebrity memorabilia: photos; a framed copy of the National Medal of Science; a reproduction of an oil portrait of him by the Australian artist Lewis Miller that makes him look "really tough"; newspaper clippings ("The Scientist vs. The Movie Star," about his run-in with Robert Redford and the Sierra Club over recombinant DNA research); postcards of Lincoln, Mendel, and Darwin; an original Salvador Dalí collage entitled *Homage to Crick and Watson*; and an old-fashioned shop sign, "Help Wanted: No Irish Need Apply," which he bought on eBay for $25. "Because people are so scared of genetic engineering, I always get a big laugh when I tell them, 'Hey, the Irish need all the help they can get.'" Ba da boom. Watson just can't resist a joke.

The photo he's looking for is not up there; right after he and Crick published their first paper in 1953, announcing the discovery of the double helix, *Fortune* featured him as one of the "Ten Top Young Scientists in US Universities." "In 1962, I made *Life* as one of the 'Red Hot 100,'" he says, chuckling, clearly exulting in the "red hot" classification rather than the quaintly reverential honors he has racked up in the intervening years. A month after the *Life* spread, he got the call from Sweden — he would share that year's Nobel for Physiology or Medicine with Crick and Wilkins. Watson, who reporters quickly learned was much better copy than his staid British colleagues, told the *New York Times:* "It is an important thing we have accomplished, but we have not done away with the common cold — which I now have."

This mischievous streak, combined with an utter lack of self-censorship, has produced a literary career as distinguished by its originality as it is renowned for its outrageous one-liners. Case in point is the opening sentence of his notorious 1968 memoir, *The Double Helix,* which reads: "I have never seen Francis Crick in a modest mood."

"Oh sure, I knew it would cause trouble," says Watson, eyes widening with unabashed glee. "I said most scientists are stupid." He pauses, furrowing his brow in an effort to quote himself accurately. "The fact is most scientists act as though they are stupid because they are wedded to some approach they can't change, meaning they are moving sideways or backwards."

Friends and enemies alike prophesied a dire fate for Watson, but all the controversy propelled the book into a huge bestseller. It didn't make him popular at Harvard, however, where he was passed over for promotion, and after a rocky tenure finally left in 1976 to head up his own venture at Cold Spring Harbor. His burgeoning fame, coupled with a tendency to be insulting, had become a sore point. "When I was there they called me the Caligula of the biology department," he recalls. "I just said there was some bad science in the biology department, and you are not supposed to say that at Harvard. But I studied at the University of Chicago, where you called crap 'crap.' It was the style of South Chicago, where there were other words, such as 'bullshit,' that were common, and a stronger word, which I won't mention."

Watson enjoys a healthy give-and-take, is bored by the pedantic, and loathes piety in all its forms. He launches into a story about an upcoming seminar in California. "It's this gathering of intellectuals," he says dryly, "so they invited Edward Wilson, Richard Dawkins, and myself. Anyway, Ed Wilson was telling someone at Cambridge that he will not come because they have invited this Danish person, Bjørn Lomborg, who has written this controversial book that states that the environmental crisis isn't all that bad. So he won't go to the meeting." Watson breaks off, assuming a look of outright horror worthy of Vincent Price. "He can't point out any facts that are wrong," he continues, "he just doesn't like his message. But Wilson has no reason to say that we need all these species. What if we only had half? Would we really disappear?"

"Unwarranted self-confidence," he concludes, referring to the undoing of many of the great minds he has worked with over the years. "Linus Pauling had it to a very high degree. How else did he spend the last forty years of his life peddling that vitamin C stuff?"

Watson still does not mince his words. Last year he published a sequel to *The Double Helix*, an even more insiderish, gossipy memoir he called *Genes, Girls, and Gamow* (excerpted in *Seed* magazine), which chronicles his dual obsession with finding RNA and The Right Woman, and managed to irritate all over again all those who had finally forgiven him for the many indiscretions in his first book. "It was just my attempt at a little humor — everyone missed it," he says ruefully.

He flips through the book, which like most of his memoirs is devoted to his endless quest to get laid, and congratulates himself on including an epic rejection scene in which the girl he loves dumps him. He stops at the photo of an attractive brunette in the epilogue. "See, in the end I found the girl," he says, referring to his wife, a Radcliffe student named Elizabeth Lewis, whom he married in 1968.

"I drop a lot of names in this chapter," he says, in that way of his that is both boastful and self-mocking. He mentions an outing on the media mogul Paul Allen's yacht, and then insists on reading aloud a bit about being invited up to the British media baron Robert Maxwell's huge suite at the Hotel National in Moscow, and how Crick worried about tagging along "until the heavyset, swarthy publisher told him that any friend of Jim Watson was Robert Maxwell's friend, too." He looks up and laughs loudly. He loves this conceit because it portrays him as more of a party animal than your run-of-the-mill Nobel laureate.

The phone rings, and Watson fields a call from a colleague who is setting up television interviews in London for the release of his new book, *DNA: The Secret of Life*, published to coincide with the fiftieth anniversary of his DNA breakthrough. He strategizes on the timing of an upcoming documentary, coolly pushes for sooner rather than later, and hangs up. His latest book trots through modern DNA research, from the double helix to the mapping of the human genome (a project he spearheaded), genetically engineered fruit, and genetically modified babies. In the process, he treads on as many politically correct toes as possible, pointing out government biases, human foibles, and social challenges to future research — and fails to say anything bad about the bell curve, guaranteeing, he says cheerfully, "that it won't get a good write-up in the *New York Review of Books*."

Watson has no patience for anything that smacks of political correctness. Leaving Harvard was "life-saving" because neither the students nor the faculty would have tolerated his fearless compulsion for telling it like it is. "The only people who believe in political correctness are academics," he says, taking up a favorite rant. "The ordinary person knows what they are saying is stupid. Largely left-wing people have seized control of the humanities and social sciences, and partially even the sciences, and it's creepy." He believes the biggest problem facing society in the next decade will be that, as genetics gets so much better, "we may discover that all humans may not be equal. That might mean that certain people will not be able to perform certain jobs," he says, courting controversy with every word. "Sure, we talk about making better plants and making better animals, but you're not allowed to talk about making better humans."

If there is one dogma Watson truly despises it's that of the Catholic Church. He likes to say that the two stupidest sentences in the English language are "love thy enemy" and "the meek shall inherit the earth." The latter, he says, is just not true. "And in the human world, if you don't have enemies you aren't doing any good. Whenever he needs cheering up he tunes in to Jay Leno, he says, "because he tells at least one joke about the Catholic Church every night."

The tape recorder clicks off with a loud snap. It's late, and the one-hour interview has stretched to almost three. Watson, at home in the limelight, could have happily talked all night. As we walk out together, he points to several elegant drawings of women by André Derain, all nudes, that decorate his office walls. "Every painting I bought before I was married was of a woman," he says, eyes widening with delight at the prospect of whipping up a little last-minute trouble. He wants to go out with a bang. He nods toward the wall by the door, where a large pinup calendar is still open to Miss December, a bodacious blonde in an alluring red number. "A gift from my son," he says proudly. "I suspect this is the only prestigious academic office in the United States with one on the wall."

Watson is enjoying himself, and at this stage, is no longer bothered by critics — if he ever was. He recently purchased a 4,000-square-foot duplex that occupies two full floors of a luxury Manhattan tower — something he might have once worried was over the top. Now he just shrugs. "People will probably just say that I'm very Darwinian."

DANIEL C. DENNETT

# The Mythical Threat of Genetic Determinism

FROM *The Chronicle of Higher Education*

IT IS TIME to set minds at ease by raising the "specter" of "genetic determinism" and banishing it once and for all. According to Stephen Jay Gould, genetic determinists believe the following: "If we are programmed to be what we are, then these traits are ineluctable. We may, at best, channel them, but we cannot change them either by will, education, or culture."

If this is genetic determinism, then we can all breathe a sigh of relief: There are no genetic determinists. I have never encountered anybody who claims that will, education, and culture cannot change many, if not all, of our genetically inherited traits. My genetic tendency to myopia is canceled by the eyeglasses I wear (but I do have to want to wear them); and many of those who would otherwise suffer from one genetic disease or another can have the symptoms postponed indefinitely by being educated about the importance of a particular diet, or by the culture-borne gift of one prescription medicine or another. If you have the gene for the disease phenylketonuria, all you have to do to avoid its undesirable effects is stop eating food containing phenylalanine. What is inevitable doesn't depend on whether determinism reigns, but on whether or not there are steps we can take, based on information we can get in time to take those steps, to avoid the foreseen harm.

There are two requirements for meaningful choice: information and a path for the information to guide. Without one, the other is useless or worse. In his excellent survey of contemporary genetics, Matt Ridley drives the point home with the poignant example of

Huntington's disease, which is "pure fatalism, undiluted by environmental variability. Good living, good medicine, healthy food, loving families, or great riches can do nothing about it." This is in sharp contrast to all the equally undesirable genetic predispositions that we can do something about. And it is for just this reason that many people who are likely, given their family tree, to have the Huntington's mutation choose not to take the simple test that would tell them with virtual certainty whether they have it. But note that if and when a path opens up, as it may in the future, for treating those who have Huntington's mutation, these same people will be first in line to take the test.

Gould and others have declared their firm opposition to "genetic determinism," but I doubt if anybody thinks our genetic endowments are infinitely revisable. It is all but impossible that I will ever give birth, thanks to my Y chromosome. I cannot change this by either will, education, or culture — at least not in my lifetime (but who knows what another century of science will make possible?). So at least for the foreseeable future, some of my genes fix some parts of my destiny without any real prospect of exemption. If that is genetic determinism, we are all genetic determinists, Gould included. Once the caricatures are set aside, what remains, at best, are honest differences of opinion about just how much intervention it would take to counteract one genetic tendency or another and, more important, whether such intervention would be justified.

These are important moral and political issues, but they often become next to impossible to discuss in a calm and reasonable way. Besides, what would be so specially bad about *genetic* determinism? Wouldn't environmental determinism be just as dreadful? Consider a parallel definition of *environmental* determinism:

"If we have been raised and educated in a particular cultural environment, then the traits imposed on us by that environment are ineluctable. We may at best channel them, but we cannot change them either by will, further education, or by adopting a different culture."

The Jesuits have often been quoted (I don't know how accurately) as saying: "Give me a child until he is seven, and I will show you the man." An exaggeration for effect, surely, but there is little doubt that early education and other major events of childhood can have a profound effect on later life. There are studies, for instance, that suggest that such dire events as being rejected by your mother in the first year of life increases your likelihood of committing a violent crime. Again, we mustn't make the mistake of equating determinism with inevitability. What we need to examine empirically — and this can vary just as dramatically in environmental settings as in genetic settings — is whether the undesirable effects, however large, can be avoided by steps we can take.

Consider the affliction known as not knowing a word of Chinese. I suffer from it, thanks entirely to environmental influences early in my childhood (my genes had nothing — nothing directly — to do with it). If I were to move to China, however, I could soon enough be "cured," with some effort on my part, though I would no doubt bear deep and unalterable signs of my deprivation, readily detectable by any native Chinese speaker, for the rest of my life. But I could certainly get good enough in Chinese to be held responsible for actions I might take under the influence of Chinese speakers I encountered.

Isn't it true that whatever isn't determined by our genes must be determined by our environment? What else is there? There's Nature and there's Nurture. Is there also some X, some further contributor to what we are? There's Chance. Luck. This extra ingredient is important but doesn't have to come from the quantum bowels of our atoms or from some distant star. It is all around us in the causeless coin-flipping of our noisy world, automatically filling in the gaps of specification left unfixed by our genes, and unfixed by salient causes in our environment. This is particularly evident in the way the trillions of connections between cells in our brains are formed. It has been recognized for years that the human genome, large as it is, is much too small to specify (in its gene recipes) all the connections that are formed between neurons. What happens is that the genes specify processes that set in motion huge population growth of neurons — many times more neurons than our brains will eventually use — and these neurons send out exploratory branches, at random (at pseudo-random, of course), and many of these happen to connect to other neurons in ways that are detectably useful (detectable by the mindless processes of brain-pruning).

These winning connections tend to survive, while the losing connections die, to be dismantled so that their parts can be recycled in the next generation of hopeful neuron growth a few days later.

This selective environment within the brain (especially within the brain of the fetus, long before it encounters the outside environment) no more specifies the final connections than the genes do; saliencies in both genes and developmental environment influence and prune the growth, but there is plenty that is left to chance.

When the human genome was recently published, and it was announced that we have "only" about 30,000 genes (by today's assumptions about how to identify and count genes), not the 100,000 genes that some experts had surmised, there was an amusing sigh of relief in the press. Whew! "We" are not just the products of our genes; "we" get to contribute all the specifications that those 70,000 genes would otherwise have "fixed" in us! And how, one might ask, are "we" to do this? Aren't we under just as much of a threat from the dread environment, nasty old Nurture with its insidious indoctrination techniques? When Nature and Nurture have done their work, will there be anything left over to be me?

Does it matter what the tradeoff is if, one way or another, our genes and our environment (including chance) divide up the spoils and "fix" our characters? Perhaps it seems that the environment is a more benign source of determination since, after all, "we can change the environment." That is true, but we can't change a person's *past* environment any more than we can change her parents, and environmental adjustments in the future can be just as vigorously addressed to undoing prior genetic constraints as prior environmental constraints. And we are now on the verge of being able to adjust the genetic future almost as readily as the environmental future.

Suppose you know that any child of yours will have a problem that can be alleviated by either an adjustment to its genes or an adjustment to its environment. There can be many valid reasons for favoring one treatment policy over another, but it is certainly not obvious that one of these options should be ruled out on moral or metaphysical grounds. Suppose, to make up an imaginary case that will probably soon be outrun by reality, you are a committed Inuit who believes life above the Arctic Circle is the only life worth living, and suppose you are told that your children will be genetically ill-equipped for living in such an environment. You can move to the tropics, where they will be fine — at the cost of giving up their environmental heritage — or you can adjust their genomes, permitting them to continue living in the Arctic world, at the cost (if it is one) of the loss of some aspect of their "natural" genetic heritage.

The issue is not about determinism, either genetic or environmental or both together; the issue is about *what we can change* whether or not our world is deterministic. A fascinating perspective on the misguided issue of genetic determinism is provided by Jared Diamond in his magnificent book *Guns, Germs, and Steel* (1997). The question Diamond poses, and largely answers, is why it is that "Western" people (Europeans or Eurasians) have conquered, colonized, and otherwise dominated "Third World" people instead of vice versa. Why didn't the human populations of the Americas or Africa, for instance, create worldwide empires by invading, killing, and enslaving Europeans? Is the answer . . . genetic? Is science showing us that the ultimate source of Western dominance is in our genes? On first encountering this question, many people — even highly sophisticated scientists — jump to the conclusion that Diamond, by merely addressing this question, must be entertaining some awful racist hypothesis about European genetic superiority. So rattled are they by this suspicion that they have a hard time taking in the fact (which he must labor mightily to drive home) that he is saying just about the opposite: The secret explanation lies not in our genes, not in human genes, but it does lie to a very large extent in genes — the genes of the plants and animals that were the wild ancestors of all the domesticated species of human agriculture.

Prison wardens have a rule of thumb: If it can happen, it will happen. What they mean is that any gap in security, any ineffective prohibition or surveillance or weakness in the barriers, will soon enough be found and exploited to the full by the prisoners. Why? The intentional stance makes it clear: The prisoners are intentional systems who are smart, resourceful, and frustrated; as such they amount to a huge supply of informed desire with lots of free time in which to explore their worlds. Their search procedure will be as good as exhaustive, and they will be able to tell the best moves from the second-best. Count on them to find whatever is there to be found.

Diamond exploits the same rule of thumb, assuming that people anywhere in the world have always been just about as smart, as thrifty, as opportunistic, as disciplined, as foresighted, as people anywhere else, and then showing that indeed people have always found what was there to be found. To a good first approximation, all the domesticable wild species have been domesticated. The reason the Eurasians got a head start on technology is because they got a head start on agriculture, and they got that because among the wild plants and animals in their vicinity 10,000 years ago were ideal candidates for domestication. There were grasses that were genetically close to superplants that could be arrived at more or less by accident, just a few mutations away from big-head, nutritious grains, and animals that because of their social nature were genetically close to herdable animals that bred easily in captivity. (Maize in the Western Hemisphere took longer to domesticate in part because it had a greater genetic distance to travel away from its wild precursor.)

And, of course, the key portion of the selection events that covered this ground, before modern agronomy, was what Darwin called "unconscious selection" — the largely unwitting and certainly uninformed bias implicit in the behavior patterns of people who had only the narrowest vision of what they were doing and why. Accidents of biogeography, and hence of environment, were the major causes, the constraints that "fixed" the opportunities of people wherever they lived. Thanks to living for millennia in close proximity to their many varieties of domesticated animals, Eurasians developed immunity to the various disease pathogens that jumped from their animal hosts to human hosts — here is a profound role played by human genes, and one confirmed beyond a shadow of a doubt — and when thanks to their technology, they were able to travel long distances and encounter other peoples, their germs did many times the damage that their guns and steel did.

What are we to say about Diamond and his thesis? Is he a dread genetic determinist, or a dread environmental determinist? He is neither, of course, for both these species of bogeyman are as mythical as werewolves. By increasing the information we have about the various causes of the constraints that limit our current opportunities, he has increased our powers to avoid what we want to avoid, prevent what we want to prevent. Knowledge of the roles of our genes, and the genes of the other species around us, is not the enemy of human freedom, but one of its best friends.

GREGG EASTERBROOK

# We're All Gonna Die!

FROM *Wired*

OMIGOD, EARTH'S CORE is about to explode, destroying the planet and everything on it! That is, unless a gigantic asteroid strikes first. Or an advanced physics experiment goes haywire, negating space-time in a runaway chain reaction. Or the sun's distant companion star, Nemesis, sends an untimely barrage of comets our way. Or . . . Not long ago, such cosmic thrills, chills, and spills were confined to comic books, sci-fi movies, and the Book of Revelation. Lately, though, they've seeped into a broader arena, filling not only late-night talk radio, where such topics don't seem particularly out of place, but also earnest TV documentaries, slick mass-market magazines, newspapers, and a growing number of purportedly nonfiction books. Everywhere you turn, pundits are predicting biblical-scale disaster. In many scenarios, mankind is the culprit, unleashing atmospheric carbon dioxide, genetically engineered organisms, or runaway nanobots to exact a bitter revenge for scientific meddling. But even if human deployment of technology proves benign, Mother Nature will assert her primacy through virulent pathogens, killer asteroids, marauding comets, exploding supernovas, and other such happenstances of mass destruction.

Fringe thinking? Hardly. Sober Ph.D.s are behind these thoughts. Citing the hazard of genetically engineered viruses, eminent astrophysicist Stephen Hawking has said, "I don't think the human race will survive the next thousand years." Martin Rees, the knighted British astronomer, agrees; he gives us a fifty-fifty chance. Serious thinkers such as Pulitzer Prize winner Laurie Garrett, author of *The Coming Plague,* and Bill Joy, who wrote *Wired*'s own 2000 article "Why the Future Doesn't Need Us," warn of techno-calamity. Stephen Petranek, editor in chief of the science monthly *Discover,* crisscrosses the world lecturing on "15 Major Risks to the World and Life as We Know It." University of Maryland arms-control scholar John Steinbruner is lobbying organizations like the American Association for the Advancement of Science and the World Medical Association to establish an international review board with the power to ban research into the Pandora's box of biomedicine.

If we're talking about doomsday — the end of human civilization — many scenarios simply don't measure up. A single nuclear bomb ignited by terrorists, for example, would be awful beyond words, but life would go on. People and machines might converge in ways that you and I would find ghastly, but from the standpoint of the future, they would probably represent an adaptation. Environmental collapse might make parts of the globe unpleasant, but considering that the biosphere has survived ice ages, it wouldn't be the final curtain. Depression, which has become ten times more prevalent in Western nations in the postwar era, might grow so widespread that vast numbers of people would refuse to get out of bed, a possibility that Petranek suggested in a doomsday talk at the Technology Entertainment Design conference in 2002. But Marcel Proust, as miserable as he was, wrote *Remembrance of Things Past* while lying in bed.

Of course, some worries are truly worrisome. Nuclear war might extinguish humanity, or at least bring an end to industrial civilization. The fact that tensions among the United States, Russia, and China are low right now is no guarantee they'll remain so. Beyond the superpowers, India and Pakistan have demonstrated nuclear capability; North Korea either has or soon will have it; Japan may go nuclear if North Korea does; Iran and other countries could join the club before long. Radiation-spewing bombs raining from the sky would, no doubt, be cataclysmic. If you're in the mood to keep yourself up at night, nuclear war remains a good subject to ponder. But reversal of the planet's magnetic field?

At a time of global unease, worst-case scenarios have a certain appeal, not unlike reality TV. And it's only natural to focus on danger; if nature hadn't programmed human beings to be wary, the species might not have gotten this far. But a little perspective is in order.

Let's review the various doomsday theories, from least threatening to most. If the end is inevitable, at least there won't be any surprises.

## 1. Laws of Probability!

Standing at the Berlin Wall in 1969, Princeton astrophysicist J. Richard Gott III used a statistical formula to predict that the barrier would last 2.66 to 24 more years. It lasted 20. Later, Gott applied the same equation to humanity and calculated, with 95 percent certainty, that it would last 205,000 to 8 million more years. His paper on the subject made it into the august British scientific journal *Nature*. Basically, Gott's formula (you will be spared the details) combines a series of estimates, then treats the result as though it was precise. Speculations about the far future have about as much chance of being spot-on as next week's weather forecast. But Gott's academic reputation won't suffer; if humanity still exists in 8.1 million years, it will be a little late to revoke his tenure.

## 2. Chemical Weapons!

Spooky-sounding, sure. And dangerous. But bombs and bullets are dangerous, too. In actual use, chemical weapons have proven no more deadly, pound for pound, than conventional explosives. In World War I, the British and German armies expended *one ton* of chemical agents per enemy fatality.

Are modern nerve agents like sarin superdeadly in a way World War I mustard gas was not? When the Aum Shinrikyo cult attacked Tokyo's subway system with that substance in 1995 — the subway being an enclosed area, ideal for chemicals — twelve people died. That was twelve too many, but a conventional bomb the same size as the cult's canisters, detonated on a packed subway, would have killed more.

During this winter's duct tape scare, I heard a Washington, D.C., radio talk-show host sternly lecture listeners to flee if "a huge cloud of poison gas" were slowly floating across the city. Noxious clouds of death may float across movie screens, but no military in the real world can create them. Wind rapidly disperses nerve agents, and sunlight breaks them down. Outdoors, a severe chemical attack

likely would be confined to a few city blocks. Some chemical inci-
dents have been horrifyingly deadly. In 1994, when a Union Car-
bide plant accidentally loosed a cloud of methyl isocyanate over
Bhopal, India, 8,000 people died, some of them twenty miles from
the site. But the source was an industrial complex, and it spewed
gas for an extended period of time, something no bomb or aircraft
could do. Another heinous event, Iraq's poison gas attack on the
Kurdish town of Halabja in 1988, killed an estimated 5,000. However, the slaughter involved dozens of Iraqi aircraft flying repeated
sorties over an undefended city. Had they dropped conventional
bombs, the toll might have been equally high.

One reason U.S. and Russian leaders agreed to destroy their
stocks of battlefield chemicals was that generals on both sides realized conventional weapons were just as deadly and easier to control. You don't want to be near VX nerve gas, but then you don't
want to be near a lunatic with a single-action Colt pistol, either.

## 3. Germ Warfare!

Like chemical agents, biological weapons have never lived up to
their billing in popular culture. Consider the 1995 medical thriller
*Outbreak,* in which a highly contagious virus takes out entire towns.
The reality is quite different. Weaponized smallpox escaped from
a Soviet laboratory in Aralsk, Kazakhstan, in 1971; three people
died, no epidemic followed. In 1979, weapons-grade anthrax got
out of a Soviet facility in Sverdlovsk (now called Ekaterinburg);
sixty-eight died, no epidemic. The loss of life was tragic, but no
greater than could have been caused by a single conventional
bomb.

In 1989, workers at a U.S. government facility near Washington
were accidentally exposed to Ebola virus. They walked around the
community and hung out with family and friends for several days
before the mistake was discovered. No one died.

The fact is, evolution has spent millions of years conditioning
mammals to resist germs. Consider the Black Plague. It was the
worst known pathogen in history, loose in a Middle Ages society
of poor public health, awful sanitation, and no antibiotics. Yet it
didn't kill off humanity. Most people who were caught in the epidemic survived. Any superbug introduced into today's Western

world would encounter top-notch public health, excellent sanitation, and an array of medicines specifically engineered to kill bioagents.

Perhaps one day some aspiring Dr. Evil will invent a bug that bypasses the immune system. Because it is possible some novel superdisease could be invented, or that existing pathogens like smallpox could be genetically altered to make them more virulent (two-thirds of those who contract natural smallpox survive), biological agents are a legitimate concern. They may turn increasingly troublesome as time passes and knowledge of biotechnology becomes harder to control, allowing individuals or small groups to cook up nasty germs as readily as they can buy guns today. But no superplague has ever come close to wiping out humanity before, and it seems unlikely to happen in the future.

## 4. Chain Reactions!

The fear that scientists tinkering with the elementary components of matter might unleash disaster has a rich and distinguished history. Before the detonation of the first atomic bomb at Trinity Site in 1945, Robert Oppenheimer worried that the unprecedented heat might spark a fusion chain reaction in the atmosphere. Physicist Hans Bethe performed calculations proving the planet wouldn't ignite, and the test went ahead.

The possibility of runaway chain reactions reemerged when scientists began deploying advanced particle accelerators, like the Cosmotron built at Long Island's Brookhaven National Labs in 1952. Some scientists worried that slamming protons into antiprotons at extremely high velocities might generate an unnatural subatomic template to which other particles would bind, collapsing matter into a void, possibly for vast distances. Panels of earnest researchers met to discuss whether high-energy physics experiments might crush the planet out of existence. They decided the risk was insignificant, but their concern was reflected in Kurt Vonnegut's 1963 novel *Cat's Cradle*, in which a researcher inadvertently creates "ice-nine," a template molecule that turns water into a solid at room temperature. When a bit of the stuff falls into the sea, all water on Earth quickly solidifies, including the water in living things.

Martin Rees, who has taken part in panels evaluating the safety of particle accelerators, has revived the idea that high-energy physics could accidentally destroy the world. In his new book, *Our Final Hour*, Rees worries that power improvements in atom smashers like Brookhaven's new Relativistic Heavy Ion Collider might make these machines capable of creating a black hole that would scarf up the globe. Ever more powerful accelerators, he fears, might create a "strangelet" of ultracompressed quarks — the smallest known units of matter — that would serve as an ice-nine *for the entire universe*, causing all matter to bind to the strangelet and disappear. Since, fundamentally, matter seems to be made of very rapidly spinning nothingness, there may be no reason why it couldn't spontaneously return to nothing.

"The present vacuum could be fragile and unstable," Rees frets in his book. A particle accelerator might cause a tiny bit of space to undergo a "phase transition" back to the primordial not-anything condition that preceded the big bang. Nothingness would expand at the speed of light, deleting everything in its path. Owing to light speed, not even advanced aliens would see the mega-destructo wave front coming. In other words, a careless Brookhaven postdoc chopsticking Chinese takeout might inadvertently destroy the cosmos.

Can ordinary people evaluate the likelihood of such an event? Not without years of graduate-level study. The only options are to believe the doomsayers or regard them in light of the fact that, in the 15 billion years since the big bang, in a universe full of starry infernos and cosmic cataclysms, their nightmares haven't come to pass so far.

## 5. Runaway Nanobots!

Eric Drexler, the father of nanotechnology, calls it "gray goo": the state of things in the wake of microscopic machines capable of breaking down matter and reassembling it into copies of themselves. Nanobots could swarm over Earth like intelligent locusts, Drexler fears, then buzz out into the cosmos devouring everything they encountered. Michael Crichton's latest novel, *Prey*, describes a last-ditch attempt by scientists to destroy such contraptions before they take over the world.

Set aside the fact that, for all the nanobot speculation you've seen (including in *Wired*), these creatures do not, technically speaking, exist. Suppose they did. As the visionary scientist Freeman Dyson pointed out in his *New York Review of Books* critique of *Prey*, not only wouldn't nanobots be able to swarm after helpless victims as they do in the novel, they'd barely be able to move at all. Laws of physics dictate that the smaller something is, the greater its drag when moving through water or air.

"The top speed of a swimmer or flyer is proportional to its length," Dyson notes. "A generous upper limit to the speed of a nanorobot flying through air or swimming through water would be a tenth of an inch per second, barely fast enough to chase a snail."

## 6. Voracious Black Holes!

A supermassive black hole roughly the weight of 3 million suns almost certainly occupies the center of the Milky Way. And smaller (actually, lighter) ones are probably wandering around in space.

If such a rogue black hole happened to find its way into the solar system, its gravitational influence would disrupt the orbits of all the planets and their moons. Earth might slingshot out of the temperate range it now occupies and into frigid reaches more familiar to Mars, or it might be pushed closer to the sun to be singed, charred, or vaporized. Worse, if a sufficiently large black hole were to pass through the globe, it might be lights-out in more ways than one. The planet would be sucked into a vortex of such intense gravity that nothing would escape. The atoms that once made up Earth would be crushed out of existence as it's currently understood.

An encounter between Earth and a black hole is astronomically, as it were, improbable. However, collisions with supermassive objects of any kind would not be survivable.

## 7. Shifting Magnetic Poles!

As Earth turns, spinning molten rock in its core generates a magnetic field that surrounds the planet. The magma hasn't stopped turning, as happens in the movie *The Core*. But magnetic effects preserved in Oregon lava flows show that the world's magnetic polarity swaps from time to time. Exactly what causes these reversals is unknown. The last one seems to have happened 16 million years ago, but some researchers speculate that Earth's polarity may change as often as every 10,000 years.

In the aftermath of such an event, a compass needle would point toward Antarctica — but it's the event itself that worries some scientists. As the magnetic poles lurch, charged bodies of lava would suddenly become repelled by areas that once attracted them, causing earthquakes and other seismic disturbances. All magnetic fields might collapse briefly, playing havoc with electronics. Earth's magnetic field repels some forms of solar and cosmic rays; if the field faltered, radiation would pound the planet's surface, possibly killing plants, animals, and people in significant numbers.

It's hard to know how scary polar shift really is, since the frequency of such events is unknown. Anyway, what can anyone do about it? Nada.

## 8. Supervolcanoes!

Pompeii (A.D. 79), Tambora (1815), Mount St. Helens (1980): Be glad you weren't picnicking nearby. These exploding mountains obliterated the countryside for miles around. Anomalies in a geologically stable world, right? Wrong. The world's most storied eruptions — Krakatau in Indonesia, for example, which caused frigid winters in Europe after it blew in 1883 — were modest by the standards of volcanic history.

Much of India sits on a basalt formation geologists call the Deccan Traps. Hundreds or thousands of huge volcanoes are believed to have erupted in this region, the cataclysms lasting many millennia and covering much of the subcontinent with molten basalt to a depth of 1,000 feet. The Deccan Traps burst forth about 65 million years ago, coincident with the dinosaurian demise. Some researchers think the meteorite usually blamed for that event struck with such violence that it cracked tectonic plates, setting in motion unimaginable seismic upheaval. In addition to 100 percent destruction within the path of the 1,000-foot tidal wave of lava, the Deccan Traps eruptions would have caused an ice age, choking global megasmog, and acid rain from hell.

The Deccan Traps were a municipal fireworks display compared with a huge Siberian basalt formation called the Siberian Traps, the product of eruptions lasting 600,000 years. Those occurred about 250 million years ago, coincident with the Permian extinction — the worst mass extinction in the fossil record.

Then there are supervolcanoes, individual eruptors of extraordinary size and power, far more potent than Krakatau. Some are geologically recent. A supervolcano called Toba exploded near Sumatra 73,000 years ago. Toba pumped 5 billion tons of sulfuric acid into the atmosphere and spewed so much sun-blocking ash that global temperatures are believed to have fallen nine degrees Fahrenheit for several years — the difference between current temperatures and those of the Pleistocene ice age. Remember the "out of Africa" theory that we're all descended from a small group of people who lived in Olduvai Gorge? They may have been the sole members of genus *Homo* to survive the supervolcano's global aftereffects.

Nobody knows what triggered the Toba eruption or how to estimate when the next supervolcano will detonate. Disturbing thought: According to the U.S. Geologic Survey, a supervolcano in Yellowstone National Park may be ripe for explosion.

## 9. Sudden Climate Change!

The world has become one degree warmer in the past century. So far, that rise hasn't hurt anyone — in fact, it may have contributed to the ever-higher crop yields that have staved off predicted Malthusian famines — but it's reasonable to expect that global temperatures will get warmer, owing at least in part to artificial greenhouse gases. Eventually the extra warmth might cease to be benign.

A more pressing worry, increasingly entertained by researchers, is a sudden climate "flip." Scientists regard fossilized oxygen isotopes as proxy measures of past atmospheric temperatures. Based on isotope levels, Russell Graham of the Denver Museum of Nature and Science has identified at least sixty-three sudden flip-flops between cold and warm trends in the last 1.6 million years — a climate flip every two millennia, on average. Note that 10,000 years have passed since the current pleasantly temperate period began, so another sudden shift is overdue.

The notion that greenhouse gases could trigger such a rapid change keeps serious scientists up at night. Ocean currents, whose dynamics are poorly understood, appear to have been central to past climate shifts. What if they suddenly started changing? Western Europe — most of which lies to the north of Maine — is nicely habitable owing to the Gulf Stream, a conveyor belt of warm water that churns past England. If global warming somehow altered the

Gulf Stream's course, the European Union might be plunged into a deep freeze even as world temperatures rise.

If the past is a guide, this could happen as rapidly as over the course of a few years. Yes, people would adapt, but their numbers might be much smaller by the time the adaptation was complete. And since scientists today have little understanding of past climate flips, it's impossible to say when the next one will start. So be prepared: Stock lots of sweaters and a few Hawaiian shirts. The weather can be tricky this time of year.

## 10. Killer Asteroids!

A collision between Earth and the gargantuan Chicxulub meteorite, which left a 186-mile-long depression at the tip of Mexico's Yucatan Peninsula, probably killed off the dinosaurs. But that was 65 million years ago. It couldn't happen again. Could it?

You bet it could. Chicxulub was only one in a long line of interplanetary boulders, or near-Earth objects, that have struck the ground. And some have arrived quite recently.

In 1908, an object 250 feet across hit Tunguska, Siberia, flattening trees for 1,000 square miles and detonating with a force estimated at ten megatons, or 700 times the power of the Hiroshima blast. Had the Tunguska rock hit Moscow or Tokyo, those cities might no longer exist. In A.D. 535, a swarm of meteorites kicked up enough debris to cause several years of cruel winters, possibly helping push Europe into the Dark Ages. Ten thousand years ago, something enormous struck the Argentine pampas, obliterating a significant chunk of the South American ecology with a force thought to be 18,000 times that of the Hiroshima bomb.

Estimates by Alan Harris of the Space Science Institute of Boulder, Colorado, suggest that 500,000 asteroids roughly the size of the Tunguska rock wander through Earth's orbit. Much spookier are asteroids big enough to cause a Chicxulub-class strike. At least 1,100 are believed to exist in Earth's general area, some capable of plunging the planet into a years-long freeze while showering the globe with doomsday rain as corrosive as battery acid. None of these killer rocks is known to be on a collision course with Earth — but then, the courses of hundreds have yet to be charted.

Can we stop an incoming asteroid? Not yet. NASA is trying to coordinate tracking of near-Earth objects but has no technology that could be used against them and no plan to build such technology. This may be unwise. As the former Microsoft technologist Nathan Myhrvold has written, "Most estimates of the mortality risk posed by asteroid impacts put it at about the same risk as flying on a commercial airliner. However, you have to remember that this is like the entire human race riding the plane."

### And What If Solar Neutrinos Reflecting off Jupiter Cause Runaway Ionospheric Decompensation?!

In 1972, John Maddox, editor emeritus of *Nature,* published a prescient book called *The Doomsday Syndrome.* In it, Maddox argues that most apocalyptic claims are dubious, inflated, or have such a low likelihood that rational people need not think about them. Worrying about nutty or improbable threats, he adds, only distracts the political system from dangers or problems that are entirely confirmed.

Thus Bill Clinton sat in the White House wringing his hands about the preposterous sci-fi thriller *The Cobra Event,* in which nearly everyone in New York City drops dead from an unstoppable supergerm, when he should have been worrying about al Qaeda, a confirmed threat to New York. Thus we fret about proliferating nanobots or instant cosmic doom when we ought to be devoting our time and energy to confirmed worries like 41 million Americans without health insurance. A high-calorie, low-exertion lifestyle is far more likely to harm you than a vagrant black hole.

The time and energy spent worrying would be more usefully applied to separating serious risks from long shots. For example, if there's a magnetic pole shift in Earth's near-term future, it's difficult to imagine what anyone might do about it. But an asteroid on an intercept course might be stopped. So perhaps NASA ought to take more seriously research into how to block a killer rock. The probability of one arriving soon might be small, but the calamity it caused would be terminal.

Yes, the world could end tomorrow. But if it doesn't, its problems will continue. It makes far more sense to focus on mundane troubles that are all too real.

GARRETT G. FAGAN

# Far-Out Television

FROM *Archaeology*

ONE CHILLY SUNDAY NIGHT, I turned on the Discovery Channel. In progress was *Mysteries of the Pyramids*, which informed me of the following startling facts. The pyramid shape is virtually inexplicable. It is a terrific mystery as to how this shape came to be used by so many different cultures from around the world (from Egypt to China to Mesoamerica). In the mid twentieth century, psychic Edgar Cayce envisioned a construction date for the three pyramids at Giza of 10,500 B.C., and a recent "scientific investigation" had confirmed Cayce's date by aligning the monuments with stars in Orion's Belt as they appeared in the sky at that time. The author of this "scientific investigation," Robert Bauval, had the final word: "You are lured into entering a quest, a system of learning and, ultimately, you will be initiated into the belief system that this pyramid represents."

While easy to dismiss, programs propagating pseudoarchaeological speculations — the mystical powers of pyramids, ancient astronauts, Atlantis's role in human development, etc. — air on an increasingly regular basis not only on the niche cable channels (Discovery, The Learning Channel [TLC], and The History Channel) but also occasionally on the networks (ABC, NBC, and especially Fox). "Hybrid" productions are also quite common, where good information is freely mixed with pseudoscience. *Mysteries of the Pyramids* offered pseudoarchaeological propositions side by side with reasonable deductions about pyramids, and the transition between the two styles was seamless. A viewer lacking previous knowledge about the sites presented or how archaeology works would not necessarily see any distinction between rational deductions drawn from observable evidence, baseless speculations, and ideologically driven pseudoscience.

There is little doubt that presenting science (and archaeology) on television is a difficult business. The slow pace of change in scientific thinking, the habitual lack of consensus among academics about details, and the often complex nature of the arguments involved place special pressures on producers. For science to work on television, the program needs to tell a story. The best stories are about people, so good science shows usually highlight the human element by focusing on a researcher or team of researchers, interposing expositions of scientific reasoning as an element of the narrative. A particularly effective format is what can be called "The Vindicated Thinker." Such a format typically presents us with a problem in science that many have found intractable. We are then introduced to the "hero" who struggles against adversity to emerge vindicated. Conflict, essential to all good storytelling, is introduced in the form of colleagues or peers who find the hero's obsession with the problem unhealthy, or who dissent from the proposed solution. "You need to find drama — that is, a story with a beginning, a middle, and an end — with a hero who surmounts various difficulties to reach a goal. Or you make the archaeologist the hero solving a puzzle," explains Chris Hale, producer of several archaeology shows for British and American television.

In the case of archaeology, there are added difficulties. The unspectacular and painstaking nature of the discipline does not make for particularly scintillating television. For how long will viewers sit through scenes of dirt-sifting amid knee-high ruins? A further problem is that archaeology deals, in essence, with dead people, who somehow have to come alive for the viewers. One solution is to use computer graphics to recreate now-ruined splendors. Such sequences are increasingly de rigueur in the genre. Other newly popular options include having actors portray figures from the past or emphasizing pragmatic considerations an audience can relate to. Michael Barnes, producer of the PBS series *Secrets of Lost Empires,* assembled teams of archaeologists and engineers to re-create spectacular achievements of ancient technology — building a pyramid, raising an obelisk, and firing a medieval trebuchet. His series kept a human focus on the teams of experts while reanimating the past with a set of ancient but immediate practical problems that demanded solutions. We know the ancients did these things, but how? "Trying something out in practice beats all the armchair talk," says Barnes.

There are other ways archaeology can be jazzed up for presentation on television. Compelling hooks emphasize the "mysteries," "secrets," and "treasures" of now-lost worlds. The interpretive nature of archaeology often guarantees conflict between various parties proposing divergent explanations for the same data, and this helps the goal of storytelling. As an archaeologist who took part in a *Secrets of Lost Empires* episode on Roman baths that aired in 2000, I can personally attest that the producer liked conflict among the team of experts, as pet theories were tested and professional reputations were put on the line. (The conflict was not manufactured — it emerged quite naturally.) It is an added bonus that much archaeology involves visits to dramatic locations like Egypt, Easter Island, and the Andes.

Unfortunately, the format favored by television archaeology perfectly suits the exponents of fringe ideas. For starters, pseudoarchaeologists uniformly present themselves as tackling some terrific mystery or secret of the past, one they claim (often incorrectly) has long baffled specialists. In "solving" this great mystery, pseudoarchaeologists love to strike the pose of the unappreciated genius and, as such, they are ideal candidates for the Vindicated Thinker format. (Even if their ideas are not actually vindicated by show's end, there is the possibility they will be someday.) There is often the promise of treasure at the end of the quest, the treasure of lost ancient knowledge that somehow will be of value for humankind. The wide-ranging nature of pseudoarchaeological speculations frequently requires visits not to one but to many exotic locations in a single show, as the "argument" jumps from Egypt to Peru to Easter Island, and so on. There is another powerful storytelling feature in this genre, one usually lacking in good archaeological television: a villain. For in many pseudoarchaeology shows, the villain is archaeology itself.

Two of the most sustained pseudoarchaeological presentations I have seen on television, representing a different order of magnitude from typical one-hour specials, are *Quest for the Lost Civilization* (aired in 1998 on TLC) and *Underworld: Flooded Kingdoms of the Ice Age* (aired in 2002, also on TLC). Each three hours in length, they are written and presented by former journalist Graham Hancock, perhaps today's highest-profile pseudoarchaeologist, and both enjoy tie-ins with Hancock's books *Heaven's Mirror* (1998) and *Underworld* (2002). The speculations proffered in these shows are so wide-ranging, self-contradictory, and just downright fluffy as to defy concise summary, but the essence of Hancock's version of ancient history appears to be as follows. By the end of the last ice age (ca. 12,000 years ago), humans had already achieved high civilization with great cities, grand monuments, and terrific navigation. Supposedly, much of this civilization was coastal and was wiped out in a cataclysmic sea-level rise as the glaciers retreated, so that the human race had to start all over again in ignorance of its glorious ice-age past. Hancock has been piecing together the clues for this nameless Lost Civilization from such sources as myths, local religious traditions, obscure mathematics, astronomy, and visits to ancient sites (many of them underwater). Since his ideas run counter to archaeological "orthodoxy" about how and when civilization arose, professional archaeologists are a great obstacle to Hancock's work, which they locate at the lunatic fringe. But, Hancock believes, the evidence is mounting. . . .

The pattern is familiar. A great mystery is set up, that of the Lost Civilization. The Vindicated Thinker, in this case Hancock, struggles against adversity to prove his dearly held notions. The locations are spectacular and come hard on the heels of one another. In *Quest*, for instance, we are taken to Egypt, Cambodia, Peru, Mexico, and France in the space of four minutes, and later, Easter Island, Japan, and the United Kingdom make an appearance. The end result of Hancock's quest will not only cure our species' amnesia but will, perhaps, uncover some wonderful but unspecified ancient knowledge about eternity. We have a chance, intones Hancock reverently, "to reach out and connect ourselves to the world of spirituality, to our ancestors." Computer graphics enhance Hancock's speculations at every turn, resolving hazy sonar images into spectacular citadels, and natural rock formations into man-made artifacts. At one point, we see 10 million square miles of the planet's surface flooded in nine seconds. To emphasize the violence of this cataclysm, which supposedly wiped out the Lost Civilization, scenes of crashing waves and ice walls shattering are intercut. In fact, the end of the last ice age saw the sea level rise 400 feet over a period of 14,000 years — an average of only 0.35 inch (0.9 cm) per year.

In *Underworld,* various scholars support Hancock's propositions of an ice age civilization in India. What the viewer is not told is that many of these experts, when their backgrounds are checked, turn out to be proponents of Hindutva, a political-religious movement that insists Indian culture is "pure" and uncontaminated by foreign influence, is far older than Western archaeologists believe, and stands at the root of all human civilization. The support of these experts is presented without any hint of the chauvinistic agenda that stands behind it. Throughout *Underworld,* opinions about archaeological matters are canvassed from geologists, geophysicists, medical doctors, local colorful characters, and a yoga teacher. Anyone's views about archaeology seem acceptable — except those of archaeologists. Professional archaeologists are onscreen largely only to agree with a specific point Hancock is making at the time. A couple of dissenters are given a few minutes to air their views, but nowhere is the chain of reasoning that underpins the standard archaeological picture of the past made apparent. The dominant impression is that of an entrenched orthodoxy resistant to all new ideas.

How can such programming be justified? I asked Charles Furneaux, commissioning editor at Channel 4 in England, which made *Underworld.* "I think this is to some extent an issue of labeling," he replied. "No one here would ever defend these shows as 'archaeological' within the strict academic meaning of the term. You could equally label them 'fun speculation' or 'provocations.'

"Hancock's credentials are well known to millions of his readers, and at the outset he makes it clear that he is a journalist and not an archaeologist," adds Furneaux. "The series is clearly identified as an authored piece and this is made transparently clear to the audience." While it is true that Hancock disarmingly states at the outset that he is not a scientist, he is consistently portrayed thereafter as a pioneering thinker challenging received wisdom in very basic ways, backed by various experts. An uninformed viewer would surely think this was cutting-edge stuff, a real challenge to the close-minded establishment. It is also dubious to claim that the absence of credentials somehow exempts a television presenter from standards of evidence and rigors of argument in a discipline that the presenter claims to be revolutionizing.

Despite Hancock's dubious claims, Furneaux believes such shows can do a service to archaeology in getting people interested in the subject when they otherwise might have been apathetic. When I shared this idea with Nicholas Flemming, a marine archaeologist who declined to appear in *Underworld,* he quipped: "It is a bit like committing a murder because that gets justice talked about. There are other ways to achieve the same end."

The appeal of pseudoarchaeology on television appears not only to be enduring but burgeoning. Many reasons can be given to account for its popularity, among them its sensational claims to be rewriting the history books from page one, the arresting mythic motifs underlying its claims (Utopian Golden Ages, the Fall from Grace, the Quest, etc.), the underdog appeal inherent in the Vindicated Thinker format, or the populist attractiveness of poking know-it-all experts in the eye.

Tom Naughton, who produced *Archaeology* for TLC in the early 1990s, points to the bottom line: "Television is not about education or providing news and information. Television is about storytelling and holding the largest audience for the longest amount of time. Programmers will do anything they can to accomplish this. Pseudoarchaeology programs are in many ways more fun to watch than programs on archaeology." Both Furneaux and Hale echo this sentiment when they complain that archaeologists tend to offer only cautious and qualified reconstructions of events during filmed interviews. This hesitancy — born of a respect for the evidence and its ability to overturn our provisional understanding of the past, and perhaps also from a fear of saying something stupid on camera that fellow academics can pick up on — is far less appealing than the simple, romantic, and unconditional stories told by pseudoarchaeologists.

Countering pseudoarchaeology on television is no easy matter. Direct criticism and point-by-point contestation can come across as nit-picking crankiness. If archaeologists are to engage the pseudoarchaeologists on the latter's home turf of popular infotainment, they need to be less coy about their achievements, less hesitant about their claims, and more willing to take risks to construct an engaging narrative. All this can be done without bastardizing the truth to the degree pseudoarchaeology does, and good examples are already at hand. The profession would do well to heed them.

In March 2002, the Discovery Channel aired *Helike: The Real Atlantis*. The title derived from the suggestion that the sudden destruction of the city of Helike in 373 B.C. stands behind Plato's Atlantis myth, composed a few decades later. The program charted the quest of Greek archaeologist Dora Katsonopolou and American physicist Steven Soter to find the lost city. Against the spectacular backdrop of the Gulf of Corinth, existing ideas about the possible location of Helike were reviewed and found wanting. Putting together evidence from a variety of sources Katsonopolou and Soter came up with a surprising theory: Helike hadn't fallen into the sea but was rather buried under sediment on land. Tension mounted as Soter sank most of his life savings into funding a core-drilling operation conducted unsuccessfully for two years. Happily, however, they both emerged vindicated — but only eventually, since they first unearthed a Roman villa and then a Mycenaean settlement before finding evidence for a Greek city of the right era to be Helike. The show combined all the elements mentioned above into an effective presentation — a human story with a beginning, a middle, and an end; obstacles faced and overcome; elaborate computer reconstructions of Helike in its heyday; spectacular locations; and the sensational "hook" of Atlantis. And it was all good archaeology.

JEFFREY M. FRIEDMAN

# A War on Obesity, Not the Obese

FROM *Science*

FOOD CONSUMES OUR INTEREST. To the hungry, it is the focal point of every thought and action. To the hundreds of millions of obese and overweight individuals, it is the siren's song, a constant temptation that must be avoided lest one suffer health consequences and stigmatization. To the non-obese, it is a source of sustenance and often pleasure. To the food and diet industries, it is big business. And to those interested in public health, it is at the root of one of the most pressing public health problems in the developed and developing world.

Alarm about obesity is sounded almost weekly in response to reports that its incidence has increased significantly over the past decade, along with a concomitant rise in its dreaded health consequences: diabetes, heart disease, and hypertension. Why is it that so many of us are obese? What has changed in such a short period of time to make us obese? Who is at fault? The food industry? The obese? Parents for not insisting that their children eat less and exercise more? The medical and scientific community for not having found a satisfactory solution? Although answers are beginning to emerge, there can be no meaningful discussion of this subject until we resist the impulse to assign blame. Nor can we hold to the simple belief that with willpower alone, one can consciously resist the allure of food and precisely control one's weight. Instead, we must look at the facts dispassionately and uninfluenced by the numerous competing interests that drive debate on this subject.

The facts are these: (i) The increasing incidence of obesity in the population is not reflected by a proportionate increase in weight;

(ii) the drive to eat is to a large extent hardwired, and differences in weight are genetically determined; and (iii) obesity can be a good thing depending on the environment in which one (or one's ancestors) finds oneself. Progress toward an understanding of the gene/environment interaction that causes obesity will require the implementation of a broad-based clinical and basic research program.

In the past decade, the incidence of obesity increased by one-third from 23.3 percent in 1991 to 30.9 percent today. During this same interval, the weight of the typical American increased by an average of approximately seven to ten pounds (depending on a person's height). Although none of us would, or should, take weight gain of this amount lightly, this difference is much smaller than the enormous variation in weight that can be observed in a cross section of the U.S. population in 2002. The fact that an incremental increase in the average weight has had a highly significant effect on the incidence of obesity is rooted in the definition of obesity. Obesity is diagnosed when weight normalized for height, or body mass index (BMI) (the weight in kilograms divided by the square of the height in meters), exceeds a defined threshold. People are said to be overweight if their BMI exceeds 25, and obese if their BMI exceeds 30. Above these BMIs, more or less, the health risks of an increased weight, or adiposity to be more precise, become increasingly evident. Because weight is distributed around a mean value in the population, an increase in the average BMI in the U.S. population from 26.7 to 28.1 (as above, seven to ten pounds) between 1991 and 2000 has led to a marked increase in the number of people with a BMI greater than 30. Thus, because obesity is defined as a threshold, a relatively small increase in average weight has had a disproportionate effect on the incidence of obesity. The effect of changes in the mean value for a trait on the frequency of disease is well established.

This analysis is not intended to minimize the importance of the fact that more than half of the U.S. population is now overweight or obese and that the environment has contributed to this public health problem. (In fact, this analysis could be viewed as good news, insofar as a relatively small achievable decrease in the average weight of our population could be of enormous public health benefit.) Rather, it is intended to highlight the fact that the change in weight attributable to any recent change in our environment, such as a change in diet or a more sedentary lifestyle, is much smaller than the enormous differences in weight, often numbering in the hundreds of pounds, that can be observed among individuals living in today's world. Thus, one might ponder why, in our current environment where almost everyone has essentially free access to calories, anyone is thin. The answer appears to reside in our genes and the way in which they interact with environmental factors.

Twin studies, adoption studies, and studies of familial aggregation confirm a major contribution of genes to the development of obesity. Indeed, the heritability of obesity is equivalent to that of height and exceeds that of many disorders for which a genetic basis is generally accepted. It is worth noting that height has also increased significantly in Western countries in the twentieth century; for example, the average U.S. Civil War soldier was five feet four inches tall. Yet, in contrast to the situation with obesity, most people readily accept the fact that genetic factors contribute to differences in stature. The critical contribution of genes to individual variation, and of environment to differences in populations, was framed by John Murray in an article reporting changes in BMI over time among students at Amherst College. Murray wrote, "In any individual's case, genetic factors play a role in determining body size but they tend to cancel out in large samples from a genetic pool, leaving levels and trends in body size that result from environmental factors." The power of the genes that regulate weight is illustrated by the following case.

Four years ago, a 200-pound nine-year-old English girl, whose legs were so large she could barely walk, was found to lack the weight-regulating hormone leptin. Treatment with leptin dramatically reduced her food intake, and that of her similarly affected cousin, to the point where they both now have body weights within the normal range for their age and live normal lives. Before leptin treatment, the younger child consumed in excess of 1,100 calories at a single meal, which is approximately half the average daily intake of an adult. With only a few leptin injections, this was reduced by 84 percent to 180 calories, the typical intake of a normal child. A number of other genes have now been causally linked to human obesity, and 5 to 6 percent of severely obese children have been shown to carry defects in known single genes. That there are likely to be other genetic forms of obesity is strongly suggested by the fact that 10 percent of morbidly obese children who do not carry mutations in known genes come from highly consanguineous (inbred) families.

In general, obesity genes encode the molecular components of the physiologic system that regulates energy balance. This system precisely matches energy intake (food) to energy expenditure to maintain constant energy stores, principally fat.

That there must be a system balancing food intake and energy expenditure is suggested by the following analysis. Over the course of a decade, a typical person consumes approximately 10 million calories, generally with only a modest change of weight. To accomplish this, food intake must precisely match energy output within 0.17 percent over that decade. This extraordinary level of precision exceeds by several orders of magnitude the ability of nutritionists to count calories and suggests that conscious factors alone are incapable of precisely regulating caloric intake.

A key element of this homeostatic system is the hormone leptin, which is produced by adipose tissue and reports nutritional information to key regulatory centers in a brain region known as the hypothalamus. Increased body fat is associated with increased levels of leptin, which then act to reduce food intake. Mutations that result in leptin deficiency are associated with massive obesity in rodents and humans. A decrease in body fat leads to a decreased level of leptin, which stimulates food intake and reduces energy expenditure. Indeed, the reduced energy expenditure observed after dieting necessitates a disproportionately low caloric intake for the stable maintenance of weight loss. It is the activation of this potent behavioral and metabolic response to weight loss that makes successful dieting so difficult.

Overall, this homeostatic system can maintain weight within a relatively narrow range. Why then are some individuals obese and others not? It appears that the intrinsic sensitivity to leptin is variable and that, in general, obese individuals are leptin resistant. Because of this, only a subset of obese people respond to leptin therapy with a significant amount of weight loss; the majority do not. The molecular basis for leptin resistance is not yet fully understood but is currently an area of intensive investigation.

The homeostatic system regulating energy balance induces a powerful drive to eat after a significant amount of weight has been lost. Feeding is a complex motivational behavior, meaning that many factors influence the likelihood that the behavior will be initiated. These factors include the unconscious urge to eat that is regulated by leptin and other hormones, the conscious desire to eat less (or more), sensory factors such as smell or taste, emotional state, and others. Key neural center(s) somehow process this diverse information. Although there is clearly cross-talk between the brain regions that produce the basic drive to eat and higher brain centers from which one might express the conscious wish to eat less, there is public disagreement about the relative potency of these often conflicting drives (as witnessed by the plethora of televised infomercials on diet plans). Those who doubt the power of basic drives, however, might note that although one can hold one's breath, this conscious act is soon overcome by the compulsion to breathe. The feeling of hunger is intense and, if not as potent as the drive to breathe, is probably no less powerful than the drive to drink when one is thirsty. This is the feeling the obese must resist after they have lost a significant amount of weight. The power of this drive is illustrated by the fact that, whatever one's motivation, dieting is generally ineffective in achieving significant weight loss over the long term. The greater the weight loss, the greater the hunger and, sooner or later for most dieters, a primal hunger trumps the conscious desire to be thin. It should be noted, however, that modest weight loss — on the order of 10 pounds — has been achieved in some studies, and weight loss of this magnitude reduces the severity of diabetes and other conditions associated with obesity. Perhaps, in advance of a weight loss strategy superior to dieting, we should reduce our expectations.

What then is the role of the environment? As noted above, the increase in weight in our population is not evenly distributed; there has been a disproportionate increase in the number of massively obese people in recent years, especially in certain ethnic groups. Mean-difference analysis of this trend reveals that in recent years the BMI of U.S. adults in the lowest percentiles has not changed nearly as much as the BMI of those in the highest percentiles. Thus, in modern times, some individuals have manifested a much greater increase of BMI than others, strongly suggesting the possibility that in our population (species) there is a subgroup that is genetically susceptible to obesity and a different subgroup that is relatively resistant.

The biologic system that regulates weight, although robust, is under intense selective pressure, and the genes that constitute it would be expected to vary depending on the environment. For people who lived in times of privation, such as hunter-gatherers, food was only sporadically available and the risk of famine was ever present. In such an environment, genes that predispose to obesity increase energy stores and provide a survival advantage in times of famine. This is the so-called "thrifty gene hypothesis" put forth by James Neel in 1962. Indeed, thrifty genes could be imagined to be genes that lead to leptin resistance, the end result of which would be the efficient retention of nutrients as adipose tissue. Consistent with this idea is the finding that obesity and an increase of plasma leptin levels, indicative of leptin resistance, are characteristic of Pima Indians living a "Western" lifestyle, whereas Pima Indians living a more "native" lifestyle remain leaner and have low leptin levels.

For people descended from the inhabitants of the Fertile Crescent or, more recently, Western societies, the risk of starvation was markedly reduced by the domestication of plants and animals and the ability to store food. But these developments also exposed those who became obese to significant health problems. In this environment, selection against obesity might be expected. Some argue that because the health consequences of obesity generally affect people beyond child-bearing age, genetic selection against obesity is not robust. However, an insightful article by Jared Diamond in 1992 suggests otherwise. Diamond pointed out that, among other things, obesity is associated with gestational diabetes, which has potentially deleterious consequences and would thus be strongly selected against. Gestational diabetes increases the risk of miscarriage and it can also lead to a cephalopelvic disproportion, an event that can have catastrophic consequences for both mother and child. Although the health complications of obesity are often not evident until later in life, it has also been shown that depriving children of the care and emotional support provided by their grandparents, especially grandmothers, has important consequences. A number of recent reports note the pivotal role of grandparents in gathering food for children and emphasize their critical role in the human social structure. In addition, increased adiposity is associated with an increased risk of predation in animals. Thus, in circumstances where the risk of starvation is reduced, one might expect genes that resist obesity and its complications to have a selective advantage. Such selection can, in principle, be quite rapid. As eloquently outlined by Jon Weiner in *The Beak of the Finch*, there is preexisting variation in all natural populations. As a consequence, natural selection can be observed in a single generation as nature weeds out the maladapted under changing environmental conditions, leaving the more highly adapted individuals to proliferate. Thus, rapid changes in population characteristics are generally the result of a gene/environment interaction.

Today, most people in Western societies have access to an abundance of food, and they lead a more sedentary lifestyle than did hunter-gatherers. However, as a species, we carry the genetic legacy of both environments. Might it be that it is the obese who carry the "hunter-gatherer" genes and the lean who carry the "Fertile Crescent" or "Western" genes? In support of this idea is the observation that populations that were historically most prone to starvation become the most obese when exposed to a Western diet and more sedentary lifestyle. This is true of the Pima Indians, Pacific Islanders, and many other high-risk populations. Thus, in modern times, obesity and leptin resistance appear to be the residue of genetic variants that were more adaptive in a previous environment. If true, this means that the root of the problem is the interaction of our genes with our environment. The lean carry genes that protect them from the consequences of obesity, whereas the obese carry genes that are atavisms of a time of nutritional privation in which they no longer live. (Some elements of this argument are an extension of the ideas in Jared Diamond's brilliant book *Guns, Germs, and Steel*.)

Once a molecular framework for the system regulating weight has been more fully developed, the next frontier will be the identification of the genes and genetic variants that cause obesity in humans. Enormous advances have been made, and the progress of the genome project will further accelerate such efforts. As more elements of this physiologic system are added, the impact of environment on their function will become better understood. An understanding of why obesity is associated with diabetes, heart disease, and hypertension is also needed. We also need to understand whether diets with different nutrient compositions have different effects on weight regulation. Still, patience is called for; scientific advances take time, and the translation of those advances into new treatments often takes even longer. The field of cancer research, for example, was invigorated by the elucidation of the molecular machinery that controls cell division. However, it was not until recently that this new molecular understanding was translated into entirely new types of therapy, such as the protein-tyrosine kinase inhibitor Gleevec, with more to come. Our approach to the obesity epidemic should be analogous: Identify the molecular components of the system that regulates body weight, define what is different about the system in lean and obese subjects, and elucidate how environmental and developmental factors alter the function of this system. Such a foundation is essential for the development of rational therapies. Substantial advances have been made, and it is a propitious time to discuss the need for a large effort aimed at understanding the biological basis of obesity.

In the meantime, a different kind of understanding is called for. Obesity is not a personal failing. In trying to lose weight, the obese are fighting a difficult battle. It is a battle against biology, a battle that only the intrepid take on and one in which only a few prevail.

ATUL GAWANDE

# *Desperate Measures*

FROM *The New Yorker*

ON NOVEMBER 28, 1942, an errant match set alight the paper
fronds of a fake electric-lit palm tree in a corner of the Cocoanut
Grove night club near Boston's theater district and started one of
the worst fires in American history. The flames caught onto the fabric decorating the ceiling, and then swept everywhere, engulfing
the place within minutes. The club was jammed with almost a thousand revelers that night. Its few exit doors were either locked or
blocked, and hundreds of people were trapped inside. Rescue
workers had to break through walls to get to them. Those with any
signs of life were sent primarily to two hospitals — Massachusetts
General Hospital and Boston City Hospital. At Boston City Hospital, doctors and nurses gave the patients the standard treatment for
their burns. At MGH, however, an iconoclastic surgeon named Oliver Cope decided to try an experiment on the victims. Francis
Daniels Moore, then a fourth-year surgical resident, was one of
only two doctors working on the emergency ward when the victims
came in. The experience, and the experiment, changed him. And,
because they did, modern medicine would never be the same.

It had been a slow night, and Moore, who was twenty-nine years
old, was up in his call room listening to a football game on the radio. At around 10:30 P.M., he heard the familiar whine of an ambu-
lance arriving outside, put on his white coat, and went to see what
was going on. Making his way to the ward, he heard another ambu-
lance arriving — then another, and another. He broke into a run.
In less than two hours, he received 114 burn victims. He described
the scene several days later in a letter to his parents:

Down the hall were streaming stretchers with burned people on them. One a young girl, with her clothing burned off, and her skin hanging like ribbons as she flailed her arms around, screaming with pain. Another a naval lieutenant who kept repeating over and over again, "I must find her. I must find her." His face and hands were the dead paper-white that only a deep third-degree burn can be, and I knew only looking at him for a moment that if he lived, in two weeks his face would be a red, unrecognizable slough. He didn't live.

Moore grabbed a syringe full of morphine and gave anyone he found alive a slug for the pain. Dozens died in those first few hours. Many succumbed from shock and overwhelming injuries. Others, some without a single burn on them, died of asphyxia, their singed throats slowly swelling closed. In all, nearly 500 people died from the fire. Of MGH's 114 patients, only 39 survived long enough to be admitted to the hospital and treated for their burns. Bodies were laid in rows along a corridor; a hospital floor was cleared for the survivors. And, at Oliver Cope's insistence, the experiment began.

The conventional treatment for severe burns was to tan the burn surface as quickly as possible, and at Boston City that is what the surgeons did. People who initially survive bad burns remain at high risk of dying from infection in the days to follow. Your skin protects you from the germs of the outside world; a burn opens the portals. Applying tannic acid to a burn was a way to create a thickened, protective cover. Patients were given morphine and soaked in a bathtub; then their blisters were cut off and the acid slowly poured on their wounds. The process was extraordinarily painful and laborious, and sometimes fatal. It also took four or five trained personnel to care for one patient. Still, it was a proved therapy, and it had been standard practice for years.

Medicine, especially surgery, is a conservative profession; a physician departs only reluctantly from the established techniques and lessons. And for good reason: The stakes, if you are wrong, are too high. Doctors are expected to adopt new treatments only with strong evidence that they will have better results. But Cope was a believer; and one of the things he believed was that tannic-acid treatment was no good.

Earlier that year, he'd been called on for advice following the Japanese bombing of Pearl Harbor. Investigators had found that the major casualties were not from blast injuries but from burns from the fires that followed. Medical personnel had been overwhelmed by the labor required by tannic-acid treatment. A day and a half after the attack, they still had not completed the initial care for victims. Many patients died waiting. Cope proposed wrapping people's burns in gauze coated with petroleum jelly, and then leaving them alone. The treatment would be far less painful, and a single doctor could care for four or five patients by himself.

To many, the notion that a thin layer of gauze smeared with Vaseline would stop infection seemed foolish. Yet on the night of the Cocoanut Grove fire — despite numerous lives at stake, intense media attention, criticism from the surgical establishment, and the fact that MGH's results could be compared very easily with Boston City's — Cope insisted on his experimental treatment. His experience with it was almost laughably slim: He had tried it just twice (once on himself). But he had found in the laboratory that blisters protected by petroleum-jelly gauze appeared to stay sterile.

Francis Moore was intelligent and ambitious; he was also, before the fire, still relatively unformed. He had received his undergraduate and medical degrees at Harvard but had distinguished himself more as a wit than as a scholar. Chronic asthma from childhood had kept him out of the war. He was no better skilled in the operating room than his fellow residents. But the Cocoanut Grove disaster became his Omaha Beach; it exposed him to a magnitude of suffering he scarcely imagined possible, leaving him with a thickened, protective hide of his own. And it gave him his first true expertise — he saw more serious burns from this one catastrophe than most surgeons do in their careers, and he came to publish dozens of papers in the field. Despite the range of his eventual influence, he called himself a burn surgeon the rest of his life.

Moore — or Franny, as everyone called him — went on to become one of the most important surgeons of the twentieth century. He discovered the chemical composition of the human body, and was a pioneer in the development of nuclear medicine. As the youngest chairman of surgery in Harvard's history, he led his department to attempt some of the most daring medical experiments ever conducted — experiments that established, among other things, organ transplantation, heart-valve surgery, and the use of hormonal therapy against breast cancer. Along the way, the line between patients and experimental subjects was blurred; his attempts to develop new procedures inevitably cost lives as well as saved them. His advances made medicine more radical, more invasive of human bodies, and more dependent upon technology. But in November of 1942 he got his first sense of what might be possible when you put aside custom and convention.

In the hours and days after the fire, Moore carried out Cope's experimental treatment. A month later, investigators from the National Research Council arrived in Boston. At Boston City Hospital, the council's final report said, some 30 percent of the initial survivors had died, most from infections and other complications of their burns. At MGH, none of the initial survivors had died from their burn wounds. Cautious experience lost. And, as Moore would always remember, experiment won.

Moore's experiments pushed medicine harder and farther than almost anyone had contemplated — and uncomfortably beyond what contemporary medical ethics would permit. Moore did not think like a surgeon; he wasn't enthralled by technique. Doing burn research with Cope, he found his strongest interest was science.

A few months after the fire, Moore took a year off from operating to do research, enrolling in nuclear physics courses and joining a biophysics lab. In that one year in the lab, he and a young chemist, Lester Tobin, invented a method of using radioactive tracers and a Geiger counter to detect abscesses that were invisible to x-rays. The technique obviated exploratory surgery for a range of diseases, anticipating the field known as nuclear medicine. Another time, on rounds one day when he was a resident, he came up with an operation to cure ulcers. He had been seeing a patient with stress-related ulcers and proposed that severing certain nerves to the stomach could cure the problem. A year later, as a new member of the MGH surgical faculty, he performed the operation and proved it — only to find that a doctor from the University of Chicago had beaten him by just a few months. The operation became the standard treatment for peptic ulcers for the next twenty-five years.

Moore's range and creativity were astounding. "The mind could simply overpower," said Steve Rosenberg, who trained under him and is now chief of surgery at the National Cancer Institute. Moore had a gravelly, throaty voice and a booming laugh, and he was by nature an actor. As an undergraduate, he had been president of both the *Harvard Lampoon* and the Hasty Pudding Theatricals. He had written plays, composed musicals, and, with his friend Alistair Cooke, once put together a show entitled "Hades! The Ladies!" which they took on the road, performing it in the Roosevelt White House. He never lost the ability to command an audience, small or large. "He had that attractive *pause*," Leroy Vandam, an anesthesiologist who worked alongside him for decades, said.

In 1945, in an experiment that made Moore's career, he tackled a seemingly inconsequential but long-unanswered question: How much water is in the human body? Even into the 1940s, nobody had a precise idea. Scientists knew that people were made of salts and minerals and water and fats and so on in some kind of balance and distribution, but they did not know much more than that or how they might find out. Moore hit upon an idea while sitting, Archimedes-like, in his bathtub: Suppose he put a drop of heavy water in the tub water, let it diffuse, and then took a drop of the tub water out. If he measured how much the deuterium had been diluted, he'd be able to calculate exactly how much water was in the tub. And he could do the same thing in a human being.

A professor he'd met while doing isotope research sent Moore a test tube of radioactive deuterium from a nuclear reactor. (Those were looser times.) Working in the small laboratory he'd been given as a junior faculty member, he injected a few milliliters of the heavy water into some rabbits, let it equilibrate over an hour, then withdrew samples of blood by vein. He measured the concentration of deuterium using a densitometer and calculated each rabbit's total body-water volume. Then he measured the amount of water in the rabbits directly, by drying them in a vacuum-desiccator. His calculations, it turned out, were perfect.

Moore began doing heavy-water studies in human beings. He tested almost anyone he could get his hands on: laboratory workers, their families, and many others. He tried other radioactive isotopes, too — of sodium, potassium, iron, phosphorus, chromium — and gradually figured out the amounts and behaviors of the essential ingredients of the human body. Then he and a growing team of researchers began taking measurements of patients with traumatic wounds, chronic illnesses, massive hemorrhages. The work paid off: Study by study, the doctors opened the black box of human chemistry.

Here are some of the things Moore discovered. That the average adult male is 55 percent water and the adult female is 50 percent. That this difference appears during puberty, when boys lose fat and girls gain it: Muscle sequesters water and fat displaces it, so that a very skinny person can be as much as 70 percent $H_2O$. That in acute illness water leaks out of blood vessels into the soft tissues, sometimes by the liter; the heart tries to adjust by increasing its rate; and the whole process is controlled by hormones in the bloodstream. That muscle protein is burned for energy, and this is why sick people waste away.

He also learned that critical illness often arose from a simple chemical imbalance — that, for instance, what often killed patients with vomiting or severe infection was a lack of potassium, and that, with carefully calibrated infusions of salt water and potassium, these deaths could be prevented. He showed that in some patients one could correct delirium, convulsions, or heart-rhythm disturbances with small amounts of magnesium, and that sick patients require particular proteins and minerals and sugars in order to survive. Through more than a decade of studies, he worked out, for disease after disease, a step-by-step process for getting people well.

The revolutionary nature of Moore's work was recognized almost immediately. In July 1948, at the age of thirty-four, he was made full professor and chairman of surgery at Harvard Medical School's Peter Bent Brigham Hospital. In 1959, he compiled his findings in what became one of the top-selling medical textbooks ever.

In dozens of ways, Moore's research simply made medicine better — more informed, more systematic, more effective. He had made a science of convalescence. Surgery, in particular, changed dramatically. Even the most elementary operations had carried substantial risk of death — the average person stood a one-in-twenty chance of dying from an appendectomy, an operation that some 7 percent of human beings require. But Moore showed surgeons how to make a range of small adjustments — from how to titrate fluids and salts to how to compensate for bleeding or shock —

that made routine surgery safer by orders of magnitude. His findings probably saved tens of thousands of lives a year — an impact on the scale of vaccines and antibiotics.

At the same time, Moore's work, by lowering the risk of invading the human body, gave doctors the confidence to try things that would once have seemed outlandish. And, in this, no one was more ambitious than Francis D. Moore.

I knew Dr. Moore as a formidable though distant presence in my surgical training. The hospital where he was chairman of surgery is my hospital — it is now called Brigham and Women's Hospital — and though he was long retired when my residency began, a few years ago he still attended weekly surgical conferences. You always shook a little when you took the podium to explain a patient's case and looked out to find him sitting in the front row. White-haired to his eyebrows, slouching a bit in his seat, his hands folded casually, he hardly appeared threatening. Yet when the time came for his comments he seemed to know everything about the science and the surgery and the saving of human lives, and he didn't hesitate to make his views known.

In an operating room, he could be chatty and make little jokes. But even when he was in his thirties, with staff sometimes decades older than he, the room was firmly in his command. He liked to whistle during cases, for example. (Some surmised that he did so to control his wheezy, asthmatic breathing — he had a poorly concealed cigarette habit that had turned his childhood lung troubles into emphysema, but he could whistle an entire Brandenburg concerto during an operation without dropping a note.) And when he did, no one made a sound.

Although he was not an especially fast or elegant surgeon, he was meticulous, methodical, and, when he needed to be, inventive. He didn't like surprises in operations, but he came prepared for them. Some surgeons arrive in the operating room with only a vague idea of what they will do. Faced with a nonroutine problem — an unusually massive hernia, say, or a colon cancer invading the spleen — they improvise, relying on instinct and experience. Moore thought hard about each case ahead of time. Once he took a patient to the operating room, he had decided on every move he wanted to make.

Like many attending surgeons, he used to browbeat the residents who operated with him. But his concern was not their technique. "He really wanted to know how you thought," Gordon Vineyard, a Boston surgeon who had trained under him, told me. "He wanted to know what you were going to do. How. Why. Especially why. He got in your head." People had put their lives in your hands, so you'd better know everything you could about them. "He'd ask you about a lab test for a patient, and, if you didn't know the result, that was a moral failure," Vineyard said. "But if you ever guessed, or made it up, you were *dead*."

"The fundamental act of medical care is assumption of responsibility . . . complete responsibility for the welfare of the patient," Moore wrote on the first page of his textbook. A good doctor, he went on, "employs any effective means available." And if there is no effective means available? Then you must try to come up with one. Death, he argued, must never be seen as acceptable. Confronted with a dying patient, he did not hesitate to consider the most outrageous proposals.

He saw, for example, scores of patients with advanced, metastatic breast cancers. The women invariably died, most within a year and in terrible pain, especially once the tumor had spread to their bones. Moore was troubled that medicine had given up on these women; it was his ethical and scientific judgment that no problem was beyond trying to solve. He established a team of physicians dedicated to treating these patients, and began to study the increasing evidence that certain hormones played a role in promoting breast cancer. In a radical 1955 experiment, Moore and a neurosurgeon named Don Matson began operating on the metastatic-breast-cancer patients to remove their pituitary gland — a whitish, peanut-size nerve center in the brain that controls the production of everything from estrogen to human growth hormone. The operation would shut down their endocrine systems and thereby, Moore hoped, eliminate a major stimulus of cancer growth.

Moore and his colleagues had no animal studies to point to, and the operation was tremendously risky: It required entering the skull above the right eye, exposing the frontal lobe of the brain, removing a portion just to get in far enough, then working over to the part of the brain behind the top of the nose where the pituitary dangled from its stalk. Of their first fifty-three patients, one lost the right half of her vision; several lost their sense of smell; three had major strokes; seven developed seizures; and one died. All had to go on medications to maintain minimum levels of essential hormones. Many physicians thought the whole project was insane.

Slowly, however, surprising results emerged. One patient was a forty-two-year-old woman with two children and a cancer of her right breast which had not only spread to her bones but formed thick red deposits all over her skin. She was bedridden and failing rapidly. After the operation, her skin tumors began melting away. She returned to a life at home, living in remission for fourteen months. In others, the rib and spinal metastases shrank, and their pains disappeared.

Many patients saw no benefits at all. Some lives were clearly made worse. But 60 percent of the women improved or survived longer, and a few were actually cured — a feat that no other medical treatment had ever accomplished. With further experimentation, Moore and another colleague, Richard Wilson, found that removing the ovaries and adrenal glands (where estrogen and other hormones are produced) had equal benefits with less risk. Other physicians demonstrated similar results, and a search began for a medicine that could provide the same hormonal effects. These efforts produced the anti-estrogen drug tamoxifen, which is now a central element in both the treatment and the prevention of breast cancer.

Moore did not hesitate to expose his patients to suffering and death if a new idea made scientific sense to him. When he became chairman of surgery, one of his earliest recruits was Dwight Harken, who, during the Second World War, had performed 134 successful operations to remove bullets and shell fragments from inside the hearts of wounded soldiers. The heart had long been regarded as impossible surgical territory. But Harken was young and brash, and, through trial and error, he accumulated more experience at opening the heart than anyone else in the world. As soon as Moore brought him in, Harken began attempting a dangerous operation called a mitral valvulotomy. The mitral valve is one of the four major valves of the heart, and ensures that blood flows forward through the pump, instead of jetting backward. In patients with mitral stenosis (which is common after rheumatic fever), the valve becomes hardened and calcified; getting blood through it can be like trying to push a river through a straw. The afflicted die by degrees, their lungs swollen with fluid, their lips turning blue, their breathing increasingly labored. In the 1920s, surgeons had attempted opening the valve in several patients by using, first, a metal punch-hole instrument, and then simply a gloved index finger to crack through. But the results were so dismal — five of seven died from complications — that the operation was abandoned.

Moore and Harken, however, took encouragement from the fact that two people had survived. Harken's initial results were hardly an improvement. Four of his first six patients died on the table. Criticism from the medical community mounted. The surgical residents at the hospital, who had to assist in these operations and care for the patients afterward, met privately with Moore to demand a moratorium on the procedure. Even Harken began to have doubts. Still, Moore was insistent that he go on, and kept the residents in line. He pushed Harken to select healthier mitral patients for surgery, arguing that the other patients might have been too ill to survive. Perhaps doing an unproved, deadly operation was permissible as a last-ditch effort for people with only days of life remaining. But Moore wanted to operate on patients who had many months ahead of them. Harken took his advice, and gradually figured out how to improve the survival rates. In 1964, he published a complete series of more than fifteen hundred of these operations. The survival rate for those with end-stage disease was 83 percent. For patients with less advanced disease, it was a stunning 99.4 percent. The era of open-heart surgery was born.

How could you be sure that a new procedure would one day prove its worth? You couldn't. During the same period, in the 1950s, Moore led a surgical team that removed people's adrenal glands in an effort to treat severe hypertension of unknown cause. Eventually, the operation had to be abandoned; the death rate was too high, and there was little evidence that the procedure helped. But Moore's confidence in his instincts was never more bitterly tested than during his department's experiments with organ transplantation.

The taking of body parts from one being for another had a long and unsavory history. In the eighteenth century, for example, a

British surgeon demonstrated that human teeth could be transplanted, and this became briefly popular in England. But it proved to be a disaster. Teeth were stolen from corpses and purchased from the poor for the gums of the gap-toothed wealthy. And though the transplanted teeth sometimes took, they also brought with them infections, including syphilis, that spread unhindered through people's jaws. In France just before the First World War, two attempts at kidney transplantation failed. Then, between the wars, there was a vogue for testicular transplants, which were thought to produce sexual "rejuvenation." In Paris between 1921 and 1926, a Russian surgeon named Serge Voronoff transplanted wedges of monkey testis into close to 1,000 men. In Kansas, by 1930, J. R. Brinkley reported having performed 16,000 glandular transplants. Laboratory studies found, however, that within days of transplantation the grafts were destroyed by the host's rejection response. Finally, the medical profession had had enough, and shut down outfits like Brinkley's. By the 1950s, when Moore got involved, the notion of taking organs from one person and putting them into another was widely regarded as disreputable.

To Moore, though, it combined the frisson of the daring and a moral attraction: He became interested in transplantation as a way to save the lives of those whose kidneys had failed. Kidney failure — from infection, shock, the slow damage of high blood pressure — had invariably fatal consequences. Unable to excrete water or the potassium and acids that build up in the body, its victims either drowned in their own fluids or were poisoned by their own blood. It was a condition more dreaded than cancer, and medicine had nothing to offer.

In 1951, Moore, then thirty-nine, and George Thorn, the medical chief of staff, put together a team under a young surgeon named David Hume to try giving these patients kidneys from dead people. Their plans were, in retrospect, woefully primitive. Despite overwhelming evidence that the immune system would reject any foreign organ, they had no strategy for stopping the rejection process. And although they had experimented with the operation in dogs, they had not yet worked out how best to hook up a new kidney. (Should you connect it to the vessels and ureter of one of the failed kidneys, or put it somewhere else? Should you keep it warm or cold until you attached it?)

Nonetheless, by 1953, ten patients had undergone the experimental operation. The results were abysmal. The first patient had been a comatose twenty-nine-year-old woman whose kidneys had shut down owing to a hemorrhage following an illegal abortion. Hume found a hospital worker who let the doctors take a kidney from a relative who had just died. The woman was unable to travel, so Hume and his team did the operation at her bedside, late at night under a gooseneck lamp, the coma providing her anesthesia. They connected the kidney to blood vessels in her arm and left the organ outside the skin under rubber sheeting. Urine began dripping out of the exposed ureter almost immediately, and the next day she awoke from her coma. Within forty-eight hours, however, rejection took over. The kidney became massively swollen, ceased producing urine, and had to be disconnected. Surprisingly, the procedure gave her just enough time for her own kidneys to recover, but she died not long after from hepatitis contracted from the blood transfusions she had required. The subsequent patients did little better. Every transplanted kidney was rejected, killing nearly every patient within weeks. The one exception, Gregorio Woloshin, was a South American physician whose new kidney somehow worked for five months before it failed.

Moore, impervious to the harsh criticisms from other physicians, took encouragement from Woloshin's case. When Hume left that year for active duty in the Navy, Moore persuaded another young surgeon, Joseph Murray, to continue the work. Murray, a plastic and burn surgeon, shared Moore's omnivorous scientific curiosity and, in his quiet and devoutly Catholic way, the same unflagging optimism. Murray transplanted kidneys into six more patients, saying a prayer for each one. Again, none survived. Fifteen patients died before the doctors accepted that there was no surgical way around the natural process of rejection.

In late 1954, the team received for treatment a twenty-two-year-old named Richard Herrick, who was deathly ill with kidney failure from scarlet fever. He was incoherent and shaken by frequent convulsions; his family was desperate. He was, in other words, like scores of patients the team had seen, except for one detail: He had an identical twin, Ronald. There were a few scattered reports suggesting that skin grafts between identical twins didn't suffer rejection. So how about a kidney transplant? Richard would be the test case. By any measure, it would be a dangerous undertaking. No one could be sure that this transplant would go any better than the others. And now the doctors would be putting a healthy young man at risk — subjecting him to both a major operation and the uncertain long-term consequences of having just one kidney.

Moore approved the operation without hesitation. A series of tests confirmed that the twins were identical. Adding to the pressure, a reporter who was at a police station when the twins arrived for one of these tests, a fingerprint examination, found out about the impending operation and made the news public. On December 23, 1954, under heavy press scrutiny, the urologist J. Hartwell Harrison opened up Ronald and removed one kidney. Moore carried it to the operating room next door, where Murray placed it in Richard's pelvis and sewed its ureter to his bladder and its artery and vein to the nearby blood vessels.

Ronald recovered with ease. More shakily, Richard, too, got better. The donated kidney survived, and there were no signs of rejection. Six months later, Richard returned home from the hospital (and soon married the nurse who had cared for him). This was the first genuinely successful transplant operation in history, and the news spread around the world.

It was, in truth, a freakish kind of victory — how many identical twins needed kidney transplants, after all? But it inspired Moore and Murray and their colleagues to persist. The next year, Moore, operating on dogs, attempted the first liver-transplantation operations ever performed. Murray and the nephrologist John Merrill began experimenting with ways to suppress the rejection response, which, it was now clear, was their primary barrier. The only known immune suppressant at that time was radiation — a lesson learned from the atom bombing of Japan. So, beginning in 1958, they subjected patients to massive doses of radiation — enough to destroy all their antibody-producing cells — before giving them a kidney transplant.

After the irradiation, each patient — even sicker now, with essentially no immune system — was moved into an operating room for its antiseptic environment and one-on-one, around-the-clock medical attention. This was, it turned out, one of the first intensive care units in the world (and a few years later Moore was among the first to convert an entire ward to function in this way). Many of the patients lived in these converted ORs for weeks. Once stabilized, they were injected with bone-marrow cells from a kidney donor to try to create an immune system that would tolerate donor tissue — the first attempted bone-marrow transplants — and then underwent the kidney transplant itself.

What followed, however, was a series of grisly deaths. The irradiation caused the first patient's blood to lose its ability to clot, and she died from bleeding a few days after the operation. Another patient, a twelve-year-old boy, died from the irradiation without even making it to the operating table. "Mortality was virtually complete," the surgeon Nicholas Tilney recounts in a forthcoming history of transplantation. "As a result, the medical, surgical, and nursing staff became increasingly doubtful about the entire enterprise. . . . Indeed, one senior medical resident in charge of the ward finally refused to involve himself any longer, telling John Merrill that he had officiated at enough murders and could not continue." In all, fourteen patients died before the irradiation approach was abandoned, in 1960.

In 1959, researchers at Tufts University Medical School had discovered a drug, 6-mercaptopurine, that could suppress the immune system in rabbits to the point that they could tolerate small injections of foreign cells. Almost immediately, a British surgeon, Roy Calne, began trying the drug in dog kidney transplants. All the animals died, but he reported one that lived forty-seven days without rejection, and Moore persuaded Calne to join his laboratory. Even before Calne arrived in Boston, Moore gave approval for Murray to perform human kidney transplants using the new drug. The first patient was operated on in April 1960, and died after four weeks. The second died after thirteen weeks.

Murray and Calne tried a modified version of the drug, called azathioprine. Ninety percent of the dogs they tried it on died within weeks. Still, in April 1961, the team went forward with a human transplant using the new drug. The patient died after five weeks. The team tried again with another patient, using a different dose regimen. That patient died, too. In the next few months, the surgeons tried four more human kidney transplants using the drug. All the patients died.

Then, in April 1962, after all those years and all those deaths, the surgeons had a startling success. Murray took a kidney from a man who had died undergoing a heart operation and put it into Melvin Doucette, a twenty-four-year-old accountant. Doucette was immediately, and permanently, put on azathioprine. He underwent two rejection crises, which the doctors managed with steroid injections. And then Doucette went home. He proved to be the first patient in the world to successfully receive a transplanted organ from an unrelated donor. Within months, the Brigham team had performed transplant operations on twenty-seven patients, nine of whom survived long-term. On May 3, 1963, *Time* put Moore, not quite fifty years old, on its cover. Murray went on to receive the Nobel Prize. Their results continued to improve. And kidney transplantation became a routine, life-saving operation around the world.

What kind of person can do this again and again — inflict suffering because of an unproved idea, a mere scientific hope? Inflicting suffering is part of any doctor's life, especially a surgeon's. We open, amputate, slice, and burn. I have put needles into screaming children who required two nurses to hold them down. I have put three-foot rubber tubes through people's noses down to their stomachs, though it made them gag and vomit and curse me in several languages. In many cases, I've found the prospect so loathsome that I looked for every excuse to avoid it. But you push ahead, because what you are doing will help the patient. You have experience and textbooks full of evidence that assure you it does. And, because it does help the patients, they forgive you; once in a while, they even thank you.

You also do it because you are part of a little community of people for whom doing such things is normal. I imagine that there's a limit to what I could tolerate doing to another person, no matter how good the evidence might be that it would help. I tell myself, for example, that I could never have been one of those Civil War surgeons who sawed limbs off unanesthetized soldiers to save their lives. Then again, I've circumcised babies without anesthesia just because that's the way everyone else does it.

Moore, however, pushed ahead not only despite suffering and death but also despite slim evidence, repeated failure, and the outright opposition of his peers. He was clearly a maverick. Renée Fox, a sociologist, and Judith Swazey, a historian, have published many studies of the murky, violent territory of surgical innovation, and they found that almost all of the surgeons involved — pioneers in heart transplantation, liver transplantation, the artificial heart — had what they called, in the title of their brilliant 1974 book, *The Courage to Fail.*

It is a generous phrase. After all, the surgeons' lives were not the ones on the line. And the portrait that emerges from Fox and Swazey's work is not entirely appealing. These surgeons were ruthless, bombastic, and unbending; they saw critics as either weak or stupid. They were self-absorbed and brutally competitive. They killed patients. They were also stunningly capable leaders: Their ability to find the positive in even the worst disasters was infectious. And they hewed to a simple moral stance. Moore used to tell the story of Cotton Mather and Zabdiel Boylston. In 1721, during a deadly smallpox epidemic in Boston, Mather, a clergyman, and Boylston, a doctor's son with no medical training, inoculated 247 healthy people, including Boylston's own son, with pus containing wild smallpox virus. On the mere basis of an account that Mather had heard from his slave about African practices, they argued that the procedure would be protective. Boston physicians were outraged. But the inoculation worked, and became a standard medical practice. To a physician like Moore, when people were going to die, the moral position was to do something, anything, however dangerous or unproved.

I asked Fox and Swazey how the surgeons they'd come to know coped with the pain and misery and disappointment to which they'd subjected so many patients. The answer varied, the researchers said. For many surgeons, the science itself — the process of reporting successes and failures in papers and at meetings, challenging critics, coming up with modifications when things haven't worked — was how they got by. Other surgeons had to steel themselves to proceed: As they scrubbed in for another try with an unproved operation, they made morbid, self-mocking jokes about the pointlessness of it all, the torture they were about to put some poor fellow through. A few spent hours with their patients and experienced tremendous anguish about what they were doing to them. One notoriously tough pioneer admitted to sleepless nights and long bouts of melancholy. You never knew "at the end of the day, or the end of the decade, or the end of the third of the century, whether what you were striving for was actually going to be anything resembling what you'd hoped it would be . . . whether what was being pawned off as treatment might, in a very real sense, be a disease in and of itself," he told the researchers. The work, he said, "was life-destroying." It was precisely their closeness to their patients that drove such surgeons to proceed. People they knew were dying and desperate, and these surgeons felt that they were the only ones in the world willing to take a chance to help them. On the other end of the spectrum were the pioneers who simply refused to interact with their patients as people. "I don't worry about them after the surgery," one famous surgeon told Swazey. "I just leave all that crap to the chaplains."

Moore was somewhere in between. Nobody saw Moore anguish about the plight of his patients, and he didn't go to their funerals. But he did not avoid his patients, either, however badly things turned out. All his life, he had an odd mixture of prickliness and charm. His enthusiasm and articulate intelligence accounted for the charm, but he was also icy in his convictions; Moore wanted to be right more than he wanted to be loved. He was the sort of man who insisted that his family have breakfast together every morning, despite a schedule that got him to work by six, the sort of man who tolerated no opposition when he'd decided that an experimental treatment should be tried.

Renée Fox pointed out something else about the surgeons she had known: A surprising number of them came to have second thoughts about what they did. Thomas Starzl, for example, was the first surgeon to attempt human liver transplants and also, many deaths later, to succeed with them. He went on to pioneer multiple-organ transplantation, though it took him years to bring the mortality rates down to acceptable levels. Then, later in his career, his courage, or bloody-mindedness, abandoned him. In a recently published interview with Fox, he told of becoming unable to take kidneys from living donors anymore. He had never had a death doing so; but he'd had to amputate the foot of one donor, a professional softball player, owing to complications, and the case haunted him. Over time, he became unable to do experimental work even on dogs.

Moore would never admit to having regrets, or doubts, not even during the most difficult years of kidney transplantation. Yet, finally, something shifted in him, too.

*

In 1963, both Moore and Starzl attempted liver transplants, in nine patients. All of them died horribly — bleeding and jaundiced and in shock from liver failure. The researchers called a temporary halt to the procedure and returned to the lab. A year later, Starzl decided to resume the operations, as did Roy Calne, who was now back in London. Instead of joining them, however, Moore denounced them, arguing that survival rates in animals had not improved enough. He never returned to the procedure or allowed others at his hospital to do so. In 1968, after Christiaan Barnard, in South Africa, performed the first successful heart-transplant operation and others around the world raced to follow him, Moore opposed the attempts as premature. He voiced similar concerns about the first efforts at implanting artificial hearts.

"Does the presence of a dying patient justify the doctor's taking any conceivable step regardless of its degree of hopelessness?" he wrote in an influential 1972 statement. "The answer to this question must be negative. . . . It raises false hopes for the patient and his family, it calls into discredit all of biomedical science, and it gives the impression that physicians and surgeons are adventurers rather than circumspect persons seeking to help the suffering and dying by the use of hopeful measures." Cautious experiments in animals were mandatory, he argued, before such measures could be tried in human beings.

His position was hard to fault. Only 12 of the first 130 liver recipients were long-term survivors. Heart transplantation had even worse results: Of the first 100 heart recipients, 98 died in less than six months. Cardiac surgeons were forced to call a moratorium on the procedure, and years passed before animal experiments were thought successful enough to justify broad-scale human trials again. Moore's doctrine became widely accepted policy. New technologies and operations for the terminally ill now require far greater levels of proof in animal trials before use in human beings is accepted. A few years ago, human gene-therapy experiments were brought to a halt by a single death. The doctrine made Moore a hated figure among animal-rights activists — they staked out his house, prompting him to buy a .38-caliber Smith & Wesson for protection. It would also have forbidden nearly every experiment he and the surgeons under him had conducted in the previous thirty

years. Moore had helped make medicine bold, unafraid, and at
times disturbingly invasive; this was suddenly the era of intensive
care, of chemotherapy and open-heart surgery and organ transplantation. But Moore himself never experimented with a major
new therapy in human beings again.

Moore changed, as did the rest of us. We no longer tolerate surgeons who proceed as he had once proceeded: We call them cow-
boys if we're being generous and monsters when someone dies. We
prefer the later Moore, the Moore whose almost blind scientific optimism was replaced by caution and carefully modulated skepti-
cism. This was the Moore who was among the few to testify before
Congress, in 1971, questioning whether Nixon's War on Cancer
had any chance of a rapid victory. As a medical adviser to NASA,
starting in the Apollo era, he raised serious, still unanswered concerns about the radiation risks that astronauts might incur from
prolonged travel in deep space. Through the 1970s, he conducted
a landmark series of studies of American surgical care and reported unexpectedly high rates of failure and complications.

This was also the humanist who, in the 1960s, wrote a handbook
entitled "The Dignity of the Patient" and required all his surgical residents to read it. He was solicitous in the care of his terminal patients. Although his bedside manner wasn't especially warm
— he remained more fatherly than familiar to patients — he expected his residents to regard all the details of patients' care as
their responsibility: not just medication doses and test results but
the easily forgotten basics of care, such as keeping weak and failing
bodies clean, lotioned, neatly dressed, turned and repositioned,
free of pain. It was among his most fiercely held criticisms of today's medicine that nurses no longer comforted or touched patients the way they used to. And as he became older he grew angry
at the possibility that he, too, could die this way — subjected to the
kind of routine medical aggressions that his own work had made
possible.

Moore was himself a terrible patient. At a surgical conference in
San Antonio, in 1994, he contracted Legionnaires' disease. He was
shaking with chills and sick with pneumonia. But he was convinced
that the hospital was more likely to kill him than the bacteria were.
He hounded the doctors mercilessly, refusing a CT scan that they

ordered and demanding that they justify even the antibiotic they prescribed.

When I was a junior resident, I cared for him during a one-night stay after he'd had a hernia repaired. He was eighty years old and questioned the doctors relentlessly about the anatomical details of his hernia, and about whether the repair was done properly. (The surgery failed, as it happened, and he had to return to get his hernia fixed again.)

He was used to thinking of himself as more or less invincible — not an unusual sentiment among doctors, but one that was bolstered by the luck he thought he'd had. During the Korean War, when he was thirty-eight, he was a surgical consultant to the MASH units. Flying along enemy lines, the Army L-5 two-seater he was in came under artillery fire and he was hit. Yet he somehow escaped critical injury, taking a bullet in his right arm, nowhere else, and without any lasting effect. When he was in his seventies, he developed abdominal pains, and a CT scan revealed a large mass in his pancreas. His doctors suspected advanced cancer, yet the mass proved to be benign. Surviving that, he only became surer that he was somehow different from everyone else.

Moore had been physically active all his life, his emphysema notwithstanding. He rode horses, hunted, fished; he acted in, and played music for, society theatre. He sailed competitively, once coming in second in the Halifax Race from Marblehead to Nova Scotia, and he kept up sailing for years after he retired from his surgical practice, in 1981, at the age of sixty-eight. He published scientific papers well into his eighties. He took up a campaign against managed care. In 1997, he and John H. McArthur, the former dean of Harvard Business School, published a landmark article proposing a national council on medical care, modeled on the Securities and Exchange Commission, to oversee the medical and financial standards of commercial health insurers. In 1988, when his wife, Laurie, to whom he had been married for fifty-three years, died in a car crash, he grieved and recovered and then got married again, to Katharyn Saltonstall, the widow of a longtime friend.

Then, in 2000, when he was eighty-seven, his family noticed that he was experiencing shortness of breath. He denied it, of course, but he got winded just walking along the sidewalk. He had to stop every few steps to make it up a flight of stairs. He couldn't continue with his regular sails on Buzzards Bay. Finally, he agreed to see his internist. Listening to Moore's heart, the physician heard a soft *whoosh* where there should have been a *lub-dub*. A weak valve was causing his heart to fail.

Once Moore accepted the diagnosis, he decided that he needed an operation. He wanted a surgeon to open his chest, put his hands inside, and make the problem go away. But his case wasn't bad enough for surgery. The doctors sent him home with prescriptions for drugs he'd have to take for the rest of his life — Lasix (a diuretic, to reduce the fluid congesting his heart and lungs) and Digoxin (a cardiac stimulant). He hated the idea of being dependent on the pills, though, and he didn't take them. His condition worsened.

One day, he couldn't catch his breath. He was admitted to the hospital, where he was given oxygen, threaded with catheters, and infused with enormous doses of diuretics that squeezed gallons of fluid from his lungs. His breathing soon stabilized and, over the next few days, became easier again. Before long, the doctors gave him a handful of prescriptions, and a lecture about taking them, and sent him home.

Thanks to the medical care he had helped create, this eighty-seven-year-old man could have expected to live five and a half more years, according to the National Center for Health Statistics. A gradual, protracted downward spiral is now taken as the norm; it was not a prospect that Moore could regard with any equanimity. His mind was still strong. But he was too frail to travel to scientific meetings. His life had become constricted; he knew he would never feel strong again.

Moore was contradictory on the subject of death. In his later years, he emphasized its inevitability and worried that science was keeping people alive too long. In a memoir, he told the story of an eighty-five-year-old woman he had taken care of who was badly burned in a fire. Her face and upper chest were gone. She was on a ventilator, and would need numerous skin grafts and weeks of care. Her odds of surviving were poor. If she did make it, she'd never eat normally or see normally again. Her husband had died. Her daughter hadn't seen her for years. A grandson came to visit just once. So when her condition began to worsen, Moore went ahead with his own decision. "We began to back off on her treatment," he wrote, and "when she complained of pain, we gave her plenty of morphine. A great plenty. By the clock. Soon, she died quietly and not in pain." He even advocated that doctors recognize euthanasia as a part of their responsibilities.

At other times, however, he insisted that death be fought, forestalled by any means necessary. In 1998, the wife of a close friend had become sick with diabetes, heart disease, and an ischemic, unremittingly painful leg. A vascular surgeon had attempted an arterial bypass to restore blood flow to the leg, but to no avail. Moore sent his son Chip, who was also a surgeon at my hospital, to see her. "There's a syringe of morphine in the drawer," she told Chip. "End this for me." He argued with her. Her vascular surgeon believed that her leg would gradually get better, he said. "He's wrong," she said. Her family needs her, he said. "I have nothing more to give them," she replied. He refused to administer the lethal shot. But when he saw his father he told him that, in his heart, he thought that she was right. The elder Moore exploded in rage that his son would even contemplate it.

About a year and a half ago, two days after Thanksgiving dinner with his son's family, and just after finishing breakfast with his wife, Katharyn, Francis Moore went to his study, shut the door, took out his old Smith & Wesson from the desk drawer, put it in his mouth, and shot himself. In his journal, which his family has kept private, he had evidently contemplated the idea on several occasions. In one entry, a few days earlier, he had written something to the effect that "today is the day," only to follow it with an entry saying that things hadn't seemed right after all. There was nothing self-pitying in what he wrote or said during his last few months. But he had kept the gun clean and hidden for no other purpose than to end his life.

"Was I surprised by his suicide? Not at all," Robert Bartlett, a Michigan surgeon he was close to, told me. Several others said the same thing. I found it unnerving to talk to Moore's surviving friends and colleagues about his decision. "If you ask any of us, we all have some kind of plan," one told me over tea and cookies in his small, neat assisted-living apartment with a red emergency button by the door. He had, he confided, a special bottle of pills in his bathroom cabinet.

There was principle in Francis Moore's death, but also some-

thing like anger. He left no note, despite having left a written legacy in every other respect. He shot himself with his wife in the
house — to hear it, to find him with his brains sprayed all over the
study. There was also, it must be said, a characteristic degree of
obliviousness of the pain he could and did cause others.

Which Moore was the better Moore? Strangely, I find that it is the
young Moore I miss — the one who would do anything to save
those who were thought beyond saving. Right now, in the intensive
care unit one floor above where I am writing, I have a patient who
is going to die. She is eighty-five years old, a grandmother. She was
healthy, on almost no medicines besides a daily aspirin, and lived
on her own without difficulty. A few days ago, however, she developed an ache in her abdomen which grew until she was rigid with
pain. In the emergency room, we discovered that she had a strangulated hernia, and, in the operating room, we saw that what had
been strangulated was a long loop of now purple and dead intes-
tine. We excised the gangrenous bowel, stapled the two ends back
together, and fixed her hernia. In the ICU, she improved rapidly.
But the gangrene had unleashed an inflammatory response that
swept through her whole body. In a process known as the adult
respiratory distress syndrome (ARDS), her lungs became stiff,
fibrous, and increasingly incapable of exchanging oxygen. Now
that we have given her as much oxygen and support as we could
and still failed to restore the $O_2$ levels in her blood to normal, her
death seems almost certain.

I spoke to her family outside the sliding glass doors of her room
and told them this. They were subdued and unhappy. "Isn't there
anything more you can do?" one of her children asked. And what I
told him was "No." Strictly speaking, though, that wasn't true. Certainly, we have done everything that the textbooks and journals say
we should do. But in an earlier era Francis Moore would have insisted that we use our science and ingenuity to come up with something else. We could try, for example, using a potent recombinant
protein called drotrecogin alfa, which was recently found to interfere with inflammation and to reduce mortality in patients suffering from systemic infections. We could try putting her in a hyper-
baric oxygen chamber. We could try liquid ventilation. We could
try injecting — I read about this once — fluid from pig lungs into

her airways. None of these therapies have been shown to work in animals, let alone in adults with ARDS. The expense would be high, and they carry serious risks of increasing her suffering or killing her outright. The recombinant protein, for instance, costs more than $10,000 and can cause severe bleeding. Colleagues and superiors would think I was mad and cruel if I used any of them. But it's not inconceivable that one of them could work, maybe even allow her to go home a week from now.

The family told me that she was clear about her wishes: If there was nothing more we could do to improve her situation, she did not want to be kept on life support. So we will do nothing more. To-night her heart will slow, then stop completely. She will go peace-fully, and that is good. She has had a long and happy life. But I can't help thinking about the costs of our caution. There would once have been a time, under Francis Moore, when we'd have been trying something, anything — and maybe even discovering some-thing new.

HORACE FREELAND JUDSON

# The Stuff of Genes

FROM *Smithsonian*

IN JANUARY 1953, deoxyribonucleic acid, DNA, was known to scientists numbered in the low hundreds. Its function puzzled, at most, a few score. The question of its three-dimensional molecular structure interested perhaps a dozen people. That structure was first visualized on the last day of February that year, at the Cavendish Laboratory, in Cambridge, England, by James Watson and Francis Crick. They published the first paper announcing the structure, the double helix, in the journal *Nature* in the issue dated April 25, 1953. They followed this up five weeks later with a second *Nature* paper on the structure's "genetical implications." The discovery was transcendent — it revealed a molecule simple, parsimonious, elegant to an extreme, and of shocking explanatory power. With these first papers it was already clear to that small world of scientists that we were on the way to understanding the nature and functioning of the material substance of genes — how hereditary characters are transmitted down the generations and how they are expressed in the development and differentiation, the building, of every organism.

So now we're marking the fiftieth anniversary of the publication of the structure of DNA. Is fifty years just a convenient round number, or does it have any significance, any resonance? As it turned out, the anniversary coincides with the closing of one era in molecular biology and the transition to a new one. This is, of course, the sequencing, nearly complete, of the human genome.

Of the founders of molecular biology, some have died: Max Delbrück, Rosalind Franklin, Jacques Monod, Linus Pauling, Max

Perutz. Francis Crick lives: Theorist and generalissimo of the golden age that followed the discovery of the structure, since the 1970s he's been at the Salk Institute, in La Jolla, California, performing a similar role in a minor key for neuroscience. A uniquely powerful intelligence, at age eighty-six he is threatened by cancer but erect of carriage, clear of mind, and serene of spirit. And Jim Watson? After the Nobel Prize, he turned administrator, building the Cold Spring Harbor Laboratory into a science and graduate training factory, and he remains a power and public gadfly of science, clumsy, intuitive, manic, fearless, often right, always awkward — in that fine French phrase, a *monstre sacré*.

DNA is a term on everyone's lips, in every day's newspapers and broadcast news accounts. (Recently, I saw a new line of modernistic furniture called "dna.") We are told relentlessly of the wonders that are beginning to flow from the sequencing and manipulation of DNA — from our detailed understanding of molecular genetics and from our growing skill at altering genes. The technologies of the gene have altered food crops, we are assured, radically increasing production. They have already changed organisms in stranger ways, so that goats and bacteria, for example, produce drugs. They may one day improve the diagnosis of various diseases. We are told that perhaps they will begin to offer cures for certain diseases. From the first years in the development of the technologies of the gene, in the early 1970s, the hope arose that genetic disorders — those caused by a single gene that is missing or defective — could be cured by supplying patients' cells with the correct gene. After more than a quarter-century of fruitless efforts and the spending of hundreds of millions of dollars, gene therapy is becoming possible, though at absurd expense and with risks not yet understood. The next hope is insight, perhaps therapies, for diseases involving multiple genes, like cancers. At the least, we are told, the technologies of the gene will allow great improvements in preventive medicine, when the reading of appropriate stretches of an individual's genome will detect genetic susceptibilities to certain ailments. Then there's talk, still sotto voce, of making improvements in human hereditary characters by adding new genes to the germ line so that they are transmitted to offspring.

I'm not writing to advertise these wonders. The consequences of the discovery of DNA's structure reach far deeper. The technologies of the gene will drive at least two great political changes, in my view. The first: We will not be able to deal equitably with what the human-genome project can do for the health and medical status of individuals without moving to universal single-payer medical coverage.

More startlingly, recall what the governor of Illinois, George Ryan, did on January 10 and 11 of this year, just as he was leaving office. He pardoned four men who had been convicted of murder and condemned to die, and commuted to prison terms every one of the other death-penalty sentences in that state. Ryan had been a defender of the death penalty. His was an act of conscience and courage. But the new evidence that the death penalty kills innocent people — this evidence comes mainly from the large number of cases in which DNA identifications have exonerated prisoners on death row. The practical, the political, and the ethical and moral arguments against the death penalty haven't changed — nor have they changed public opinion much in the last half century. But now for the first time we have irrefutable scientific proof that the death penalty frequently produces irremediable injustices.

My real interest, though, is in another consequence, of a vast and entirely different order. First, a bit of history: In the summer of 1945, Vannevar Bush, an electrical engineer who was formerly vice president of MIT and director of the wartime Office of Scientific Research and Development — where he oversaw the improvement of radar, the mass production of penicillin and sulfa drugs, and the development of the atomic bomb — delivered to President Harry Truman a report titled *Science: The Endless Frontier*. Bush believed that basic research is technology's feedstock and was convinced that after the war the federal government would have to continue to organize and pay for scientific research and the training of new scientists. He called for and predicted the astonishing, high-exponential growth of the enterprise of the sciences that we have all witnessed.

The uniquely compelling part of Bush's proposal, though, was not just its assertion that basic research would produce practical payout, but that we cannot predict how long that will take, and from what particular lines. What a license for curiosity! Biologists, molecular, cell, and biomedical biologists, have been living that ideology ever since. The promise of medical advances has generated ever larger budgets.

Now, however, molecular biology is beginning to tell us things that reach beyond the practical, beyond the political, to touch and reshape our understanding of who we are and how we came to be. Since the dawn of human time, every culture has searched for stories of origins — of the universe, of the solar system, of life, of species, of humans, of language, of civilization. With gathering momentum for more than a century, science is telling better, more comprehensive, more grounded, more verifiable stories of origins.

Molecular biology has one of the greatest origin stories as yet untold — and it is now unfolding through the coming together, the fusing, of the two separate kinds of questions that have been fundamental in the century and a half since Mendel and Darwin. Biologists have always pursued questions of *how,* and these are about physiology — how creatures reproduce, eat, run around. Increasingly since the discovery of the structure of DNA, these questions have been approached in terms of the functioning of genes. Biologists have also been occupied with questions of *why.* These are about the ways we, and all other creatures, have come to have the traits and behaviors we do. "Why" questions in biology are about adaptation or extinction, the changes in species in the struggle for existence across the immense depths of geological time: In other words, they are about evolution.

Recall that not just the human genome has been sequenced, but bacteria and roundworm and fruit fly and mouse and soon chimpanzee and more, genomes over the entire range of living creatures. Now we can match these genomes up, in a new science, comparative genomics, which is just beginning to yield in full detail the fusion of genetics and evolution — of how and why. One quick example: Molecular biologists in England and Germany recently discovered a DNA sequence or gene that appears essential to the human ability to use language. The sequence controls the action of a cascade of genes, affecting several functions. (Nobody said this was going to be simple.) Some members of a large family are afflicted with a single mutation in this DNA sequence that severely limits their ability to use words, to learn and employ normal syntax. Chimpanzees have that gene sequence, too — but it is slightly different from that in humans. In such discoveries lie what my friends the biologists ought to be advertising. Here are the transcendent answers we will thrill to. Darwin said it, in the poignant last paragraph of *The Origin of Species:* "There is grandeur in this view of life." Here is the triumph of the scientific worldview.

GEOFFREY NUNBERG

# The Bloody Crossroads of Grammar and Politics

FROM *The New York Times*

IS THERE A GRAMMATICAL error in the following sentence? "Toni Morrison's genius enables her to create novels that arise from and express the injustices African Americans have endured."

The answer is no, according to the Educational Testing Service, which included the item on the preliminary College Board exams given on October 15 of last year. But Kevin Keegan, a high-school journalism teacher from Silver Spring, Maryland, protested that a number of grammar books assert that it is incorrect to use a pronoun with a possessive antecedent like "Toni Morrison's": that is, unless the pronoun is itself a possessive, as in "Toni Morrison's fans adore her books."

After months of exchanges with the tenacious Mr. Keegan, the College Board finally agreed to adjust the scores of students who had marked the underlined pronoun "her" as incorrect.

That's only fair. When you're asking students to pick out errors of grammar, you ought to make sure you haven't included anything that might bring the grammarati out of the woodwork.

But some read the test item as the token of a wider malaise. "Talk about standards," wrote David Skinner, a columnist at the conservative *Weekly Standard*. Not only had the example sentence been "proven to contain an error of grammar," but the sentence's celebration of Ms. Morrison, a "mediocre contemporary author," betrayed the "faddish, racialist, wishful thinking that our educational institutions should be guarding against."

It was revealing how easily Mr. Skinner's indignation encompassed both the grammatical and cultural implications of the sentence. In recent decades, the defense of usage standards has become a flagship issue for the cultural right: The people who are most vociferous about grammatical correctness tend to be those most dismissive of the political variety. Along the way, though, grammatical correctness itself is becoming a strangely arbitrary notion.

Take the rule about pronouns and possessives that Mr. Keegan cited in his challenge to the testing service. Unlike the hoary shibboleths about the split infinitive or beginning sentences with "but," this one is a relative newcomer, which seems to have surfaced in grammar books only in the 1960s. Wilson Follett endorsed it in his 1966 *Modern American Usage,* and it was then picked up by a number of other usage writers, including Jacques Barzun and John Simon.

The assumption behind the rule is that a pronoun has to be of the same part of speech as its antecedent. Since possessives are adjectives, the reasoning goes, they can't be followed by pronouns, even if the resulting sentence is perfectly clear.

If you accept that logic, you'll eschew sentences like "Napoleon's fame preceded him" (rewrite as "His fame preceded Napoleon"). In fact you'll have to take a red pencil to just about all of the great works of English literature, starting with Shakespeare and the King James Bible ("And Joseph's master took him, and put him into the prison"). The construction shows up in Dickens and Thackeray, not to mention H. W. Fowler's *Modern English Usage* and Strunk and White's *Elements of Style.* ("The writer's colleagues . . . have greatly helped him in the preparation of his manuscript.") And it's pervasive not just in the *New York Times* and *The New Yorker,* but in the pages of the *Weekly Standard,* not excluding Mr. Skinner's own column. ("It may be Bush's utter lack of self-doubt that his detractors hate most about him.")

The ubiquity of those examples ought to put us on our guard: Maybe the English language knows something that the usage writers don't. In fact the rule in question is a perfect example of muddy grammatical thinking. For one thing, possessives like "Mary's" aren't adjectives; they're what linguists call determiner phrases. (If you doubt that, try substituting "Mary's" for the adjective "happy" in sentences like "The child looks happy" or "We saw only healthy and happy children.")

And if a nonpossessive pronoun can't have a possessive antecedent, logic should dictate that things can't work the other way
around, either: If you're going to throw out "Hamlet's mother
loved him," then why accept "Hamlet loved his mother"? That's an
awful lot to throw over the side in the name of consistency.

But that's what "correct grammar" often comes down to nowa-
days. It has been taken over by cultists who learned everything they
needed to know about grammar in ninth grade, and who have
turned the enterprise into an insider's game of gotcha! For those
purposes, the more obscure and unintuitive the rule, the better.

Pity the poor writers who come at grammar armed only with
common sense and a knowledge of what English writers have done
in the past — they're liable to be busted for violating ordinances
they couldn't possibly have been aware of.

Not all modern usage writers take doctrinaire views of grammar,
whatever their politics. But the politicization of usage contributes
to its trivialization, and tends to vitiate it as an exercise in intellectual discrimination. The more vehemently people insist on uphold-
ing standards in general, the less need there is to justify them in the
particular. For many, usage standards boil down to the unques-
tioned truths of "traditional grammar," even if some of the traditions turn out to be only a few decades old.

Take the way Mr. Skinner asserted that the College Board examination sentence was "proven to contain an error of grammar" in
the way you might talk about a document being proven to be
a forgery: It's as if the rules of grammar were mysterious dicta
handed down from long-forgotten sages.

The English conservative writer Roger Scruton has described the
controversies over usage as merely a special case of the debate between conservative and liberal views of politics. But until fifty years
ago, nobody talked about "conservative" and "liberal" positions on
usage, and usage writers were drawn from both sides of the aisle.

Even today, it would be silly to claim that conservatives actually
care more deeply about usage standards than liberals do, much less
that they write more clearly or correctly. In language as elsewhere,
it isn't as if vices are less prevalent among the people who de-
nounce them most energetically.

But people who have reservations about the program of the cultural right often find themselves in an uneasy position when the
discussion turns to usage. How do you defend the distinction be-

tween "disinterested" and "uninterested" without suggesting that its disappearance is a harbinger of the decline of the West?

Not that the cultural left is blameless in this. Some of the usage reforms they championed have been widely adopted, and society is the better for it. There aren't a lot of male executives around who still refer to their secretaries as "my girl."

But many of the locutions and usage rules that have recently been proposed in the name of social justice are as much insider codes as the arcane strictures of the grammar cultists. They're exercises in moral fastidiousness that no one really expects will catch on generally.

To younger writers, today's discussions of usage often may seem to be less about winning consensus than about winning points. It's no wonder they tend to regard the whole business with a weary indifference. Whatever — will this be on the test?

MIKE O'CONNOR

# Ask the Bird Folks

FROM *The Cape Codder*

Dear Bird Folks:

I'm the maid of honor at my sister's spring wedding. Before I do too much planning I need to know if it is really okay to throw rice. Some people tell me rice is bad for birds and that we should throw birdseed instead. You are probably going to tell me to throw birdseed since that is what you sell, but is rice really that bad?

Kathy, Eastham

Hold on, Kathy. Are you saying I would purposely mislead you in order to sell more birdseed? Well, I am ashamed. I'm not ashamed of you, but of myself for not thinking about doing that before. What a great idea. If this economy doesn't improve soon, I just might answer every question with "Buy birdseed," no matter what the question is. But right now you'll have to settle for the truth. Or at least the truth the way I see it.

Here's the theory on wedding rice. After the happy wedding couple leaves the church, all their so-called friends show their love by pelting them with fists full of hard, uncooked rice. If they somehow are able to survive that, the couple drives away and all the tossed rice is left lying, unguarded, on the sidewalk. Soon a flock of innocent birds arrives to chow down whatever rice didn't get lodged in the target victims' eyes, ears, and underwear. Once ingested, the harmless rice quickly expands to fifty thousand times its original size, causing the birds to inflate to roughly the size of the *Hindenburg*. Not only is this rapid expansion bad for the birds, but the birds' massive bodies can block traffic for hours.

This might come as a surprise to many, but rice is a grain. Grains, seeds, and nuts are what birds have survived on for centuries. Any rice farmer will tell you that he could only hope that birds would overinflate upon stealing his rice. The rice that thousands of birds are eating in rice fields every day is the same rice that you buy in the store or that you throw at weddings. Rice, cooked or uncooked, will not bother the birds. But I wouldn't throw cooked rice at your sister's wedding without checking with her first or without letting it cool.

If I were you, Kathy, I still would throw birdseed. Birdseed is trendy. Plus, the local birds will enjoy it more than rice, and if you bought a whole pound of seed from me, I would finally pass Bill Gates on the *Forbes* richest-man-in-the-world list. Wouldn't he be surprised!

While I'm on the subject of myths, here are a few more about birds' eating habits that never seem to go away.

Peanut butter will cause birds to choke. Yeah, sure, maybe, if you were to hold a bird's mouth open and jam a shovelful of peanut butter in it. We would choke too if we did the same. Birds, like us, eat proportionately the right amount of food at one time. A chickadee takes only a tiny bit of peanut butter at once, the proper amount for a bird of that size to handle without choking. Birds, just like people, do occasionally have trouble and do indeed choke on food, but that is rare. The bad thing for birds is not eating peanut butter but trying to Heimlich each other with their wings. They can't seem to get the grip right.

This last food myth is as wacky as the rice thing. Rumor has it if you toss a gull an Alka-Seltzer, the gull will eat it and explode. That's right, explode. Listen, Alka-Seltzer might taste bad but it's not dynamite, it's medicine. Do you really think that they would sell it to people if it made things explode? All it does is produce gas, and most creatures have a way of ridding themselves of gas, with only small explosions. Like many birds, gulls are masters at regurgitating things that don't agree with them. Even though they may not explode, I'm certainly not recommending that anyone give gulls Alka-Seltzer. Although, after what I've seen gulls eat, they probably wouldn't mind a few tablets.

Dear Bird Folks:

Why is it that birds sleep standing on one leg? My aunt has a pigeon and we were wondering about this behavior. Also, how are they able to hear so well without the benefit of large ears, like mammals have? I hope I'm not cheating by asking two questions at once, but I would really appreciate the answers to both.

Emily, Tupper Lake, N.Y.

It's okay, Emily. There is no reason to feel guilty. I don't mind two questions. Wondering about birds' hearing is an excellent question. It is odd that birds can hear so well, yet they don't appear to have any ears. As for the standing-on-one-leg thing, I knew it was only a matter of time before that question came up. I might as well answer it now and get it out of the way.

Birds do indeed have excellent hearing, but like most things in nature, there are different designs for different creatures. Most land mammals have a significant outer ear, or pinna (look it up, it's a real word). Pinna size is not as important as many people think. If having large external ears allowed mammals to hear more, then Ross Perot would be listening to us right now. And elephants would have the best hearing of all, when, in fact, elephants' hearing is relatively poor. Their huge external ears are mostly used for cooling, defense posturing, and, in the case of Dumbo, flying.

Bird hearing is at best complicated and at worst boring to read about. I'm afraid that if I got into too many details even the people in prison would put down this newspaper and go back to staring at the walls. You should know, however, that it is the internal ear that is the key to birds' hearing. The ear openings of birds are larger than you might think; it's just that they are covered up with feathers. In many cases, the feathers are arranged to help channel sounds into the bird's ear.

For years keen-eared birds have been used as "watchdogs." Before the days of radar, trained parrots were used in the military. The parrots could hear the hum of distant enemy planes and would squawk a warning long before the planes could be heard by humans. Pigeons also have excellent hearing. (So your aunt had better be careful what she says about hers.) Homing pigeons apparently use audio clues (ocean sounds, waterfalls, etc.) as well as visual clues when finding their way back to the roost.

Now for your first question. Man, I would love to have a doughnut for every time someone came screaming to me about seeing a bird with a leg missing. It is very common for birds to tuck up one leg, yet no one knows for sure why. The most common theory is to conserve heat. As you know, most birds have bare-naked legs, without the covering of insulating feathers. By tucking up one leg, at least one leg can be insulated from the cold. Go to any Cape beach parking lot in the winter time and you'll see dozens of one-legged gulls looking like they are ready to tip over.

That insulation idea makes sense, until you think about flamingos living in the zillion-degree heat of Florida. Why would they ever need to keep warm? The thought here is that they lift up one foot simply to give it a chance to dry out. Standing in wet mud 24/7 has to be tough on a new manicure. Another theory is that by standing on one leg, flamingos change their silhouette and thus are able to fool predators into thinking they are some kind of odd vegetation. Although I think the real problem in trying to hide is not the number of legs but the bright, fluorescent pink feathers. But what do I know?

Why would your aunt's pigeon living in a warm cozy house still sleep on one leg? It probably stems from its wild heritage. Either that or your aunt's house has some serious draft issues.


Dear Bird Folks:
Last week, while I was taking my pug dog, Mitzi, for a walk on Marconi Beach in Wellfleet, I came upon a group of nerdy bird watchers. They were all standing in the middle of the parking lot, staring at what appeared to be a mockingbird. I was told that the bird was a "scissor-tailed flycatcher." I didn't have any binoculars with me, but the bird didn't seem to be anything special to me. What was all the excitement about?
Toby, Wellfleet

Hold on Toby. You were spending the day at the beach, with Mitzi the pug dog (who I'm sure was wearing a sweater), and you are calling a group of highly informed bird watchers "nerdy"? Let me tell you something about my fellow bird watchers. They are by far one of the most respectable . . . Wait, I know the group you are talking about. You are right, they are kind of nerdy. Forget what I said and no offense to Mitzi or her sweater.

It's too bad you didn't have binoculars with you, because scissor-tailed flycatchers are really good-looking birds. I'd be the first to admit that many birds are dull and difficult to identify. Some birds

are so similar, they are impossible to figure out. You know, like trying to tell the difference between a pug dog and a squashed loaf of
bread. But there are plenty of birds, like puffins and roadrunners,
that will even catch the eye of non-nerdy people. The scissor-tailed
flycatcher is one of those exciting, eye-catching birds. With its
pink-colored sides and elegant nine-inch-long tail, the scissor-tailed
flycatcher is a striking bird.

Mostly found in Texas, Oklahoma, and whatever that state is
above Oklahoma, scissor-tailed flycatchers are indeed a rare bird in
these parts. They are more at home flying across the open western
grasslands than the dunes of Cape Cod.

What makes this bird so fun to watch is that it is a highly active
bird and sits out in the open for all to see. Perched high on an exposed branch or post, the scissor-tailed flycatcher watches with its
keen eyes, looking for the slightest movement. Once it spots an in-
sect, it zips out after it, flashing its rose-colored sides and opening
and closing its showy tail, like, of all things, a pair of scissors.

Beetles, grasshoppers, bees, and wasps appear to be this bird's favorite food. That's right, these birds actually eat bees and wasps
and seem to enjoy them. That should really freak out those people
who worry about birds eating wedding rice.

After the breeding season, scissor-tailed flycatchers form massive
communal roosts, in which many hundreds of these splendid birds
may be seen in a single tree. In the morning the entire flock heads
off in different directions, creating what must be a spectacular
sight. Seeing a sight like that would almost make it worth going to
Texas. I said almost.

After the breeding season, a few flycatchers go to Florida, but
most of them head to Central America, where they have cleaner
elections. What this lone bird is doing so far from its normal range
is anybody's guess. It could have gotten pushed here by a storm, or
its migration instincts may have somehow failed. Or perhaps it won
five million dollars in the lottery and used it as a down payment for
a one-bedroom cottage in South Wellfleet. Whatever the reason,
scissor-tailed flycatchers do have a habit of showing up on Cape
Cod every few years.

It is now the first week of December and the bird that you saw has
been in the same spot (the parking lot at Marconi Beach) for
nearly three weeks. I have no idea how long the bird will stay there,

if it will finally get a clue and head south, or even if it will survive. But if I were you, I would grab yourself some binoculars and go back and take a good look at this flashy-looking bird, while you still can.

Scissor-tailed flycatchers are warm-weather birds, and they really aren't made to handle our Cape Cod climate. However, it is not up to us to interfere. But if seeing it out there in the cold concerns you, Toby, you can always let it borrow one of Mitzi's sweaters.

Dear Bird Folks:

At a party the other night we were discussing our favorite birds. Then we started wondering what your favorite bird is. We all took a guess and my job was to write to you to find out if any of us were right. So, what is your favorite bird?

Monica, Barnstable

Really, Monica. You went to a party and played "name your favorite bird"? Talk about living life on the edge. I hope your kids don't find out what you do at night. Did the cops come by to break it up?

My favorite bird, eh? I feel like I'm being interviewed for a teen magazine. Fine, I'll tell you my favorite bird. Maybe next week I'll tell you my favorite movie, my sign, and my turn-ons and turnoffs.

As far as I'm concerned, all birds are great. They all have fascinating behavior, incredible survival skills, and diversified beauty. Yet only one bird has all the best qualities wrapped up in one package. The black-capped chickadee is by far the best bird ever invented. I know, right now there are millions of readers (or at least dozens) screaming, "Chickadee, no way!" To which I reply, "Way." Chickadees have it all. Let us count the ways.

First of all they are so stinkin' cute. Many birds have flashier colors, but with the fancy colors comes a snotty attitude. The beautiful spring warblers can't be bothered coming close enough for us to appreciate their colors. They zip about high in the treetops and couldn't care less that we must suffer permanent neck damage staring straight up for hours, hoping for a glimpse. Meanwhile, the inquisitive little chickadee will come to the branch just above your head, or it will even land on your head if you are pleasant enough.

For a guy who makes a living selling bird stuff, chickadees are the perfect bird. They eat just about every type of seed. They love suet.

They nest in birdhouses and they come to birdbaths. Chickadees alone could put my kids through college, if, for some reason, one of my kids was accepted to a college.

Over the course of a year, Cape Cod is visited by close to 350 different species of birds, but very few can claim that they are here year-round. Some birds (and many people) hate the heat, hate the cold, or hate the crowds. To them the grass is always greener. But our chickadees are with us 24/7/365. They are able to deal with the hot, crowded summers and have learned to adapt to the freezing, boring winters. And never once do they complain about either. How many of you can say that?

Speaking of complaining, have you ever heard anyone complain about a chickadee? Has its sweet little song ever woken you up at 6:00 A.M, or have they ever taken a bite out of anything in your garden? Have they ever made a mess on your boat or drilled holes in the side of your house or charged anything on your credit card without your permission? I'm telling you these birds are perfect.

I know there are plenty of cardinal fans screaming that cardinals are the best. Please. Cardinals are a bunch of sissies. Think about it. Any time there is a predator around, crows, jays, and chickadees are the first ones to sound the alarm. Meanwhile the cardinals are nowhere to be seen. Most of the time they don't even show up at our feeders until it's almost dark.

Keep in mind, Monica, that I enjoy all birds. Just because I think chickadees are the best (and they are), that doesn't mean other birds aren't wonderful. Okay, maybe calling cardinals a bunch of sissies was a little strong, but I'm still not taking it back.

Before I sign off, here are a few other birds that are popular but have one thing or another that keeps them from being my favorite:

Loon: Handsome in the summer, boring brown in the winter.
Hummingbird: Way, way, way too hyper. They need to chill.
Titmouse: A totally embarrassing name.
Catbird: It would be a good choice if it didn't have the "C" word in its name.
Great blue heron: Reminds me too much of Florida.
Peacock: Tries too hard.
Parrots: Forget it. If you want something that talks back, get a teenager.
Falcon: No way, they eat chickadees.

Pelican: Needs to work on its breath.

Sandpiper: Too confusing.

Bluebird: State bird of New York. The Yankees live in New York.

Crows: Very sad. The way they dress reminds me of the late Johnny Cash.

Ducks: They seem so distant.

Piping plover: Hogs the headlines.

Goldfinch: Way too much molting.

Blue jay: On the edge. They could snap any minute.

Baltimore oriole: Hello. It's from Baltimore.


Dear Bird Folks:

I have had a great blue heron feeding at a bog near my house. Now everything is frozen, yet the heron is there every day. I'm worried. What can I do to help it?

Olivia, Marstons Mills

Okay, Olivia. Here's what to do. Get yourself a very long extension cord and a hairdryer. If you start now your bog should be ice free by July.

The good news is that large birds like herons can last a few days without a good meal. The bad news is that good meals for herons have gotten very hard to find this winter. Herons never eat their vegetables, no matter how hungry they are. They eat live creatures, mostly fish. With the endless amount of snow and ice that we've had this season, fish-eating birds are having a tough time.

Great blue herons live and breed just about anywhere in the normal United States, and most of Canada. When the cold weather arrives, the herons head south. A few come to Cape Cod, where the winters usually aren't too bad. Most of these herons are either inexperienced young birds or lost adult males too stubborn to ask for directions south. Spending the winter here has its advantages, and I'm not talking about the free off-season parking in Provincetown. Herons are able to avoid the dangers of migration, plus they can be one of the earliest to arrive on the breeding grounds.

However, there is a risk with staying this far north. Yes, our winters are often mild and pleasant. Then there is this winter, the winter that never ends. Snow, ice, and cold are not kind to birds, and I'd bet many herons won't be booking a visit to Cape Cod next year.

Herons have one good thing in their favor. They are excellent hunters and are total opportunists. When the fish are frozen out, they'll eat other things, including crustaceans, mice, voles, and small birds. One hungry heron was seen chowing down a litter of feral kittens. I know, I know, I too was upset to read about the herons eating small birds.

Herons also have one odd behavior that is not in their favor. In the winter they seem to choose and defend a favorite fishing hole. When these areas become frozen solid, some herons don't seem to catch on and often will stand over a frozen stream for days waiting for the fish to return. Boy, talk about stubborn.

A winter like this could hit herons hard, and we could lose a lot of them. Now before you become even more obsessed with the potential loss of your heron, try to remember that these birds are predators. Many smaller creatures are happy to see them check out (remember those poor baby birds).

Years ago, when I was soft and wimpy like you, I would go out and chop holes in the ice so the herons could find fish. The herons loved me, but the fish hated me. I would get hundreds of wet and smelly letters from angry minnow widows. Other creatures hated me too. A healthy heron has few natural enemies, but a weakened heron could easily become lunch for a long list of other predators that are also struggling this winter.

Unlike the bobwhites that we talked about last week, herons are not trapped here. Soon they will leave and return to their breeding grounds. By and large, the heron population is doing fine. Great blue herons often nest in the dead trees created by beaver ponds. Recent changes in trapping laws have increased the beaver population, thus creating acres of potential nesting areas for herons.

Don't be too upset about your heron, Olivia. The water should open up soon, and it will be fine. If the water doesn't open up, well . . . you could always try that hairdryer option.

PEGGY ORENSTEIN

# Where Have All the Lisas Gone?

FROM *The New York Times Magazine*

ACCORDING TO THE OFFICIAL Popular Baby Names Web site, the name we are considering for our daughter, to be born later this summer, was in the Top 200 for her sex last year. It was less popular than Molly but more so than Abby. This has me worried. It seems perched at a precarious point from which it could, without warning, rocket into overuse. Witness Chloe, which has shot from 184 to 24 since 1991. Call out the name in your local Gymboree, and four little heads will whip around.

Popular Baby Names, which is operated by the Social Security Administration, ranks the 1,000 most common boys' and girls' names since 1900 (www.ssa.gov/OACT/babynames/). You can also look up a specific name and track its status over time (an activity that, I warn you, is an Internet addict's sinkhole). The site, started seven years ago, was initially the side project of a government actuary named Michael Shackleford. Michael reigned as the No. 1 boys' name for thirty-five years beginning in 1964, after about a decade of duking it out with David and Robert. It was unseated by Jacob in 1999.

Shackleford grew up, with no small amount of bitterness, in a multiple-Michael world. He hoped that by publishing the list, parents-to-be would see that his name (and other common names) were shopworn and choose something more original. (Shackleford, incidentally, quit the Social Security Administration in 2000 and moved to Las Vegas, where he has become a gambling consultant known as the Wizard of Odds. His own children are named Melanie, No. 88, and Aidan, No. 63.)

Perennials like Michael or Sarah are not, to my mind, the nub of the issue. They don't explain why so many people seeking more adventurous names seem to hit upon the same ones. Why did I recently receive birth announcements from three couples who had never met, who lived as distant from one another as Maine, Minnesota, and California, yet who had all named their sons Leo? How to account for the sudden spate of Natalies?

I am not so smug as to think myself immune to first-name zeitgeist. A few years ago, I developed a sudden affection for Julia, which now hovers at 31, and then for Hannah, which is No. 3. Although I have never personally met a Madison (2), I have watched friends seduced by the seeming novelty of Alyssa (12), Olivia (10), and Dylan (24 among boys), only to discover that their children are destined to spend life with the initials of their last names appended to their first.

While my husband doesn't seem concerned — at least judged by the excessive eye rolling when I bring up another contender — I've trawled the Social Security site for clues to the potential future of "our" name. I've sifted through message boards on pregnancy sites to see if it has cropped up among other moms-to-be. I've checked a site that polls users to determine a name's image based on continuums of ambition, attractiveness, and athleticism. I've even looked on the Kabalarian Philosophy site, which, using a supposed mathematical principle, analyzes the "power" hidden in more than 500,000 names. None of that, however, explained what I really want to know: how a particular name becomes popular and whether it's inevitable, like it or not, that my husband and I will choose the next Kayla (19).

Pamela Redmond Satran and Linda Rosenkrantz have built their empire on the backs of people like me. Their eight books, including the classic *Beyond Jennifer & Jason, Madison & Montana*, have sold more than a million copies; a new volume, the pared-down and pointedly titled *Cool Names*, will be published next month. Like *Jennifer & Jason*, it is part advice manual, part pop sociology text. Avoiding the deadly (and useless) dictionary format, it divides names into sections. There's the safe Hot Cool (Polly, Harry); the famous Cool Cool (Charlize, Keanu); the retro Pre-Cool Cool (Beata, Lazarus); and the New Cool, which encompasses, among other things, constellations (Elara, Orion). The express purpose is to help jittery parents-to-be separate current favorites from what's about to break big from what the daring among them can pioneer.

The duo read the baby-name tea leaves of preschool class lists, maternity wards, and birth announcements. They also consult the Social Security site, though Satran warns of a critical glitch: It doesn't combine alternative spellings. In 1998, for instance, Kaitlyn was way down at 36. But if you totted up the Katelyns, Caitlins, Caitlyns, Kaitlins, Katelynns, Katlyns, Kaitlynns, Katelins, Caitlynns, Katlins, Katlynns, and Kaytlyns, that name would have easily bested the No. 1-ranked Emily. Like any kind of forecasting, though, from predicting cargo pants to recognizing that we're about to have an orange moment, picking the next Grace (15) is as much art as science. "We look at all the lists," Satran says. "We look at movie stars' names and what they're naming their children. We look at names that cut across several trends at once. But after that, it's just instinct."

Satran and Rosenkrantz have a pretty solid record of prognosticating, particularly on groups of names. They sounded the alarm on the use of places (Paris, Sierra, Asia) as first names in 1988, years before that trend slid from mainstream to cliché. A friend named her daughter London, Satran remembers, which caught her attention. A short time later, she heard about a baby boy named after a Pennsylvania town. She then met a Holland and heard about a Dakota. Those encounters dovetailed with an uptick of androgynous names for girls. By the time Alec Baldwin and Kim Basinger named their daughter Ireland, Satran and Rosenkrantz knew that place names were firmly on the map.

Names weren't always subject to fashion. About half of all boys in Raleigh Colony were named John, Thomas, or William, and more than half of newborn girls in the Massachusetts Bay Colony were named Mary, Elizabeth, or Sarah. Even in the twentieth century, John, William, James, and Robert were, in some combination, the top three names for boys for more than fifty years. Among girls, Mary held on to No. 1 for forty-six years, when it was supplanted for six years by Linda, fought its way back for another nine, then succumbed to the juggernaut of Lisa.

These days, even a popular name isn't especially prevalent: Though the name was ranked fourth, there were only about 16,300 Emmas born last year. Sell-by dates are shorter too, at least for girls. Only three of today's Top 10 names (Sarah, Samantha, and Ashley) survived since 1990.

With boys — well, there's Michael. Parents continue to be more conventional with their sons, more conscious of tradition and generational continuity. Girls' names are more likely to be chosen for style and beauty. That makes them both more interesting to track and more vulnerable to sounding passé, the human equivalent of bragging about your new pashmina.

The Harvard sociologist Stanley Lieberson first bumped up against the fashion quotient of names in the 1960s. Believing they were bucking convention, he and his wife named their eldest daughter Rebecca, only to discover a few years later that she was part of a pack. How had that happened? The marketplace, after all, has no interest in what we name our children; no corporation profits if you choose Kaylee over Megan. That makes names one of the rare measures of collective taste.

Lieberson, the author of *A Matter of Taste: How Names, Fashions and Culture Change,* insists that names generally rise and fall independent of larger cultural or historical events. Consider the resurgence of biblical names. "They came back like gangbusters in the late twentieth century," Lieberson says. "There was speculation that it was related to a resurgence of religion. But people who use Old Testament names are, if anything, less religious in their behavior than those who don't. It's just fashion."

Naming styles, Lieberson says, are usually variations on what came before, moving forward predictably, the way lapels get wider and wider until they reach a peak and switch direction. He calls this "the ratchet effect." Take Old Testament names. In 1916, Ruth, for no obvious reason, was the only one to crack the Top 20 for girls. After it crested, it was replaced by Judith in 1940, then Deborah in 1950. By the late 1980s, there were three Old Testament names among the top slots: Rachel, Sarah, and Rebecca. Now it's Hannah, Abigail, and Sarah, with Leah (90 and holding) as the only potential replacement. Perhaps after a hundred years, girls' biblical names have ratcheted as far as they can go.

Sometimes, Lieberson explains, rather than a concept, it's just a sound that catches hold: the "a" at the end of girls' names (Emma, Hannah, Mia, Anna), or the hard "k" at the beginning (Kylie, Kaylee, Caitlin, Courtney). That breakthrough sound undulates outward, in a kind of jazz riff, gradually mutating. So the "djeh" sound in Jennifer begat Jenna and Jessica, but Jennifer also begat Heather and Amber, which share its suffix. (Before Jennifer, the only commonly used "er" name was Esther, which was never a favorite.) Those names went on to spawn waves of their own. African-American parents, who are more likely than other groups to invent names for their daughters — again, less often for their sons — recently became enamored with "meek": Jameeka, Camika, Mikayla. (Remember the legendary three "meeks" of the Tennessee Lady Vols basketball team — Tamika Catchings, Chamique Holdsclaw, Semeka Randall?)

But why does "a" or "djeh" or "meek" appeal in the first place? Why not the "th" in Ethel and Thelma (or Ruth!) or the final "s" in Gladys and Lois? That's harder to explain. "My speculation would be that a sound like the final 'a,' which did not used to be particularly popular, probably broke through as a variation on some existing name," Lieberson says, "and then it developed its own life."

That's not to say that external forces are irrelevant. Race clearly influences naming. So does class, especially among whites. Lieberson found that highly educated mothers are more likely to give daughters names that connote strength (Elizabeth or Catherine as opposed to Tiffany or Crystal). Yet, when it comes to boys, the trend reverses, with the more bookish moms going for Julian over Chuck.

That's the problem with the Popular Baby Names site: With no nuance, no dissection by demographic, it can get you only so far. For instance, Satran and Rosenkrantz recently polled upscale nursery schools in Manhattan and Berkeley, California. Among that crowd, Charlottes (206) and Rubys (210) ran rampant, but it was a desert for Savannahs (40).

After a couple of hours of my relentless quizzing, Satran (whose own children are named Rory, Joseph, and Owen) suggested that some people become a tad obsessed by their quest for originality. While it may evoke a particular theoretical profile (Bambi, anyone?), there is no definitive evidence that a name affects an individual child's popularity, mental health, or achievement level. "There are people who want to sell the idea that your name is your destiny," Satran says. "Names aren't your destiny any more than your shoes are." She pauses, then adds, "Well, OK, maybe your shoes are your destiny."

On the other hand, when she recently advised a friend that Maya was becoming overexposed, it made no difference. Sometimes people fall in love with a name and don't want to believe it's played out. Or they're comforted by something that's a touch more common — not everyone wants to be a trendsetter, not even those who say they do.

"There's this ideal," Satran says, "not just in names but other things that have to do with style, that you should make a personal statement. But the fact is that most people are not that adventurous. They say they want individual style but they pick their furniture at Pottery Barn. So if you tell them you're going to name your child Matilda, they'll say, 'That's awful.' But if you say Sophia or Lily or any of the names that I'm totally sick of, they'll say, 'That's such a beautiful name.'"

Even pros like Satran and Rosenkrantz are occasionally blindsided by a name, as when Trinity leapfrogged to 74 after the release of *The Matrix*. Popular culture is an oft-cited launching pad for naming fads — soap operas most famously (Kayla, Hunter, Caleb, and Ashley all zoomed upward after star turns on daytime dramas). Still, the effect is not as direct as it may seem. Buffy, despite a fanatic cult devotion to the vampire slayer, has not breached the Top 1,000 (although Willow has been climbing modestly since 1998). Aaliyah surged after the singer's death, but Diana barely budged after the Princess of Wales died.

A closer look finds that Trinity was already on the upswing, from 951 in 1993 to 555 five years later. "Riding the curve," as Lieberson calls it, is often the true explanation behind a pop-name phenomenon. A name (or a sound sequence) is in the air, albeit marginally so; because of that, it's used for a character or happens to be that of a high-profile performer (like Jada, 78). That, in turn, catapults the name forward, seemingly out of nowhere.

Bringing us back to the improbable popularity of Madison: It first hit the Top 1,000 in the 1980s and it was, unlike Trinity, probably a pure media event originating in the film *Splash*. Recall that, while struggling to choose a name, Daryl Hannah's mermaid strolls onto a certain Manhattan street, et voilà.

Still, Madison? No. 2? How in the name of good taste did that happen? Satran points to a confluence of trends: Madison came along at a time when place names and surnames (McKenzie, Morgan) as first names were hot, as well as the related androgynous

names for girls (Taylor, Sydney) and the Ralph Lauren, faux
horsey-set names (Peyton, Kendall). Then there's Lieberson's pho-
netic wave theory. In this case, Madeline (56) may have begun to
grow tired while Madison sounded just a little fresher. So when
Madison finally sinks, who will replace her?

On a hunch, I typed another New York place name into the Popular Baby Names site: Brooklyn. Sure enough, it has vaulted from
755 to 155 since 1991. Then I tried expanding in a different direction on the sound chain from Madeline and discovered that
Adeline was inching up as well. Given those trends, it would not be
as random as it would appear if, a few years from now, Adelaide and
Portland, two seemingly unrelated names, were both in the Top 10.

Now I was getting somewhere. A few nights later, I saw a film that
took place around 1900, a mother lode of contemporary names
for both sexes. One character was Annabelle. That sounded jaunty.
I liked it. But what was its appeal? Then I recalled the current popularity of the Isabella/Isabel/Isabelle chain (14, 84, 112) not to
mention Anna (20) and Ella (92). Lovely names all, but they've
been done. That made me suspicious. As it turned out, Annabelle
was rising with a bullet (from 984 to 330 in seven years, while
Annabella went from 963 to 722 in just one). The following week I
spied it monogrammed on a sleeping bag in the Pottery Barn Kids
catalog. Annabelle was off my list.

Michael aside, overuse usually spells the end of a name, at least
for a while. Names also lose luster when they become tied to a particular era. If you really want to ensure your baby girl will be unique
among her peers, name her Barbara, Nancy, Karen, or Susan. Or
Peggy. Those sound like the names of middle-aged women because
— guess what? — they are.

But names are often resurrected when the generation that bears
them dies out. Although our mothers may joke that the play group
made up of Max, Rose, Sam, and Sophie sounds like the roster
of a convalescent home, contemporary parents find those names
charming. Doubtless, today's Brittany will name her daughter
Delores.

Or maybe she'll call her Remember. Satran claims that the next
big trend will be word names. Colors, for example (she just heard
of a baby Cerulean), or words that resonate with the parents' values
or professions like Integrity or Story. "There's been a street-level

thing happening for a while with names like Destiny and Genesis," she says. "They weren't mainstream, but they were there. The tipping point came when Christie Brinkley, who is very visible, named her daughter Sailor because she and her husband liked to sail. Parents are increasingly looking for names that are different and also looking for names with personal meaning. Word names are a natural place to go. It's virgin territory. Our grandchildren will have names we don't even think of as names now."

Satran expects to see a fad in heroes' last names as first names (Monet, Koufax) as well as futuristic or Asian-sounding names borrowed from video games (Vyce, Ajuki). Among African-American parents, she says, the coming thing will be idiosyncratic punctuation accelerated by the singer India.Arie and the singer Brandy, who recently named her daughter Sy'rai.

Which brings me back to the name we are considering for our daughter. We're not, as it turns out, willing to saddle her with something as outré as Minerva. And Zazie or Tallulah are just trying too hard. Our name, as the experts would predict, is a sideways hop rather than a radical leap from names that have recently been stylish. So yes, it could take off. Still, it's a little softer, a little more free-spirited than its precursors, not the sort of name you'd imagine for a future Wall Street gunner. But that suits me fine: I ditched the East Coast fifteen years ago for the sunny iconoclasm of Northern California and a life that has become far less conventional than I once imagined. I want my daughter's name, and, I suppose, her life, to reflect that.

I hesitantly asked Satran's opinion, realizing that, like the mother of Maya, I might refuse to heed it. Had we accidentally picked the next Zoe? "Nope," she said. "I think you're safe."

So what is it? I can only respond with Satran's parting piece of advice: "Don't tell anyone the name before the baby is born. Do you really need to know about the girl with that name someone hated in fourth grade?"

She's right. Besides, I don't want to start a trend.

VIRGINIA POSTREL

# The Design of Your Life

FROM *Men's Journal*

THOSE OLD SCI-FI MOVIES were wrong. The twenty-first century
doesn't look at all the way they said it would. We citizens of the
future aren't wearing conformist jumpsuits, living in utilitarian
high-rises, or getting our food in the form of dreary-looking pills.
On the contrary, we are demanding and creating a stimulating,
diverse, and strikingly well-designed world. We like our vacuum
cleaners and mobile phones to sparkle, our backpacks and laptops
to express our personalities. We expect trees and careful landscap-
ing in our parking lots, peaked roofs and decorative facades on our
supermarkets, and auto dealerships as swoopy and stylish as the
cars they sell.

"Design is everywhere, and everywhere is now designed," says
David Brown, a design consultant and the former president of Art
Center College of Design in Pasadena, California. And it all happened so fast. It wasn't that long ago that Apple's iMac turned the
personal computer from a utilitarian, putty-colored box into curvy
eye candy — blueberry, strawberry, tangerine, grape, lime. Trans-
lucent jewel tones spread to staplers and surge protectors and mi-
crowaves — even American Express cards.

Since then everything around us has been getting a much
needed face-lift. Volkswagen reinvented the Beetle. Karim Rashid
reinvented the ordinary trash can. Oxo reinvented the potato
peeler (thus proving that people will happily pay an extra five
bucks for a kitchen tool if it looks and feels better). Even toilet
brushes have become design objects, with something for every personality. The handle on Phillippe Starck's sleek Excalibur brush
looks ready for a duel, while Alessi's Merdolino brush (designed by
Stefano Giovannoni) sprouts like a bright green cartoon plant.

When Target introduced a line of housewares developed by architect-designer Michael Graves, few customers had ever heard of Graves. But his playful toaster quickly became the chain's most talked about, and most expensive, model. Target increased the number of Graves offerings to more than five hundred and is still adding more Graves items — and more designers.

"We're seeing design creep into everything," says Chicago industrial designer Mark Dziersk. The 1990s were the decade of distribution. Wal-Mart set the standard for low-cost, hyperefficient retailing, while the Internet made everything available everywhere. You can live in a small town and still buy stylish goods. "I see 2000–2010 as the decade of design," says Dziersk.

This trend doesn't mean that a particular style has triumphed or that we're necessarily living in a period of unprecedented creativity. It doesn't mean everything is now beautiful, or that people agree on basic standards of taste. Instead of a single dominant standard, we see aesthetic fluidity. Diversity and choice, not uniformity and consistency, are our new ideals. The holy grail of modern product designers is mass customization, not mass production. "Mass production offered millions of one thing to everybody," writes Bruce Sterling, the science fiction author and design champion, in *Metropolis*. "Mass customization offers millions of different models to one guy."

Ours is a pluralist age in which different styles can coexist, as long as they please the individuals who choose them. You don't just buy blue jeans anymore; you customize, picking the exact wash and cut you like best. If you don't like the look of the nearest Starbucks, the company gives you choices. "You can go three stores down to a different Starbucks and say, — I like this better. I just feel better here,'" explains a Starbucks executive. And once you're there you don't order just a cup of coffee. You navigate a long menu of customized combinations, including different beans, styles, and flavors.

All this choice required technological and business innovations, but the shift expresses deeper cultural changes, too. The extension of liberal individualism — the primacy of self-definition over hierarchy and inherited, group-determined status — has altered our aesthetic universe. Try as they may, official tastemakers no longer determine the "right way" to look. The issue is no longer what style is used but rather that style is used, consciously and conscientiously, even in areas where function used to stand alone.

Not that other values have gone away. We may crave a barbecue grill that looks like a piece of sculpture, but we still want it to work well. We get pleasure from the bright colors of Nalgene's plastic water bottles, but we also appreciate their indestructibility. We continue to care about cost, comfort, and convenience. It's just that on the margin aesthetics matters more than before.

Designers themselves are finally abandoning the modernist idea of the one best way and embracing the pleasures of personalization. "Good design is not about the perfect thing anymore but about helping a lot of different people build their own personal identities," says David Kelley, the founder of the IDEO design firm (which designed the look and feel of your iPod).

This attitude marks a huge shift. Designers and other cultural opinion leaders used to believe that a single aesthetic standard was right — that style was a manifestation of truth, virtue, even sanity. What if someone didn't like the way Bauhaus architect Walter Gropius had arranged the furniture in the new Harvard dorm he designed? "Then they are a neurotic," Gropius replied.

Today's modernists don't talk like that. They emphasize pleasure and personality. "Instead of finding a style and adhering to its tenets, modern design allows you to grapple with your own ideas about how you want to live," says Lara Hedberg, publisher of *Dwell,* the architecture and interiors magazine for "nice modernists." Just because you like austere high-tech lighting and a chrome and glass coffee table doesn't mean you can't have a comfortable upholstered chair.

The current design revolution recognizes that sensory experience is as valid a part of our nature as our capacity to speak or to reason. "We are by nature — by deep, biological nature — visual, tactile creatures," says David Brown. The objects we desire don't need any other justification for pleasing our visual, tactile, emotional nature, as long as they make life more enjoyable.

Those prophets who forecast a sterile, uniform future got it wrong because they imagined a society shaped by impersonal laws of history and technology, divorced from individuality, pleasure, and imagination. But "form follows emotion" has supplanted "form follows function" as the defining mantra of the day, along with "I'd like a grande mint mocha Frappuccino."

JONATHAN RAUCH

# Caring for Your Introvert

FROM *The Atlantic Monthly*

DO YOU KNOW someone who needs hours alone every day? Who loves quiet conversations about feelings or ideas, and can give a dynamite presentation to a big audience, but seems awkward in groups and maladroit at small talk? Who has to be dragged to parties and then needs the rest of the day to recuperate? Who growls or scowls or grunts or winces when accosted with pleasantries by people who are just trying to be nice?

If so, do you tell this person he is "too serious," or ask if he is okay? Regard him as aloof, arrogant, rude? Redouble your efforts to draw him out?

If you answered yes to these questions, chances are that you have an introvert on your hands — and that you aren't caring for him properly. Science has learned a good deal in recent years about the habits and requirements of introverts. It has even learned, by means of brain scans, that introverts process information differently from other people (I am not making this up). If you are behind the curve on this important matter, be reassured that you are not alone. Introverts may be common, but they are also among the most misunderstood and aggrieved groups in America, possibly the world.

I know. My name is Jonathan, and I am an introvert.

Oh, for years I denied it. After all, I have good social skills. I am not morose or misanthropic. Usually. I am far from shy. I love long conversations that explore intimate thoughts or passionate interests. But at last I have self-identified and come out to my friends and colleagues. In doing so, I have found myself liberated from any number of damaging misconceptions and stereotypes. Now I am here to tell you what you need to know in order to respond sensitively and supportively to your own introverted family members, friends, and colleagues. Remember, someone you know, respect, and interact with every day is an introvert, and you are probably driving this person nuts. It pays to learn the warning signs.

*What is introversion?* In its modern sense, the concept goes back to the 1920s and the psychologist Carl Jung. Today it is a mainstay of personality tests, including the widely used Myers-Briggs Type Indicator. Introverts are not necessarily shy. Shy people are anxious or frightened or self-excoriating in social settings; introverts generally are not. Introverts are also not misanthropic, though some of us do go along with Sartre as far as to say "Hell is other people at breakfast." Rather, introverts are people who find other people tiring.

Extroverts are energized by people, and wilt or fade when alone. They often seem bored by themselves, in both senses of the expression. Leave an extrovert alone for two minutes and he will reach for his cell phone. In contrast, after an hour or two of being socially "on," we introverts need to turn off and recharge. My own formula is roughly two hours alone for every hour of socializing. This isn't antisocial. It isn't a sign of depression. It does not call for medication. For introverts, to be alone with our thoughts is as restorative as sleeping, as nourishing as eating. Our motto: "I'm OK, you're OK — in small doses."

*How many people are introverts?* I performed exhaustive research on this question, in the form of a quick Google search. The answer: about 25 percent. Or: just under half. Or — my favorite — "a minority in the regular population but a majority in the gifted population."

*Are introverts misunderstood?* Wildly. That, it appears, is our lot in life. "It is very difficult for an extrovert to understand an introvert," write the education experts Jill D. Burruss and Lisa Kaenzig. (They are also the source of the quotation in the previous paragraph.) Extroverts are easy for introverts to understand, because extroverts spend so much of their time working out who they are in voluble, and frequently inescapable, interaction with other people. They are as inscrutable as puppy dogs. But the street does not run both ways. Extroverts have little or no grasp of introversion. They assume that company, especially their own, is always welcome. They cannot imagine why someone would need to be alone; indeed, they often take umbrage at the suggestion. As often as I have tried to explain the matter to extroverts, I have never sensed that any of them really understood. They listen for a moment and then go back to barking and yipping.

*Are introverts oppressed?* I would have to say so. For one thing, extroverts are overrepresented in politics, a profession in which only the garrulous are really comfortable. Look at George W. Bush. Look at Bill Clinton. They seem to come fully to life only around other people. To think of the few introverts who did rise to the top in politics — Calvin Coolidge, Richard Nixon — is merely to drive home the point. With the possible exception of Ronald Reagan, whose fabled aloofness and privateness were probably signs of a deep introverted streak (many actors, I've read, are introverts, and many introverts, when socializing, feel like actors), introverts are not considered "naturals" in politics.

Extroverts therefore dominate public life. This is a pity. If we introverts ran the world, it would no doubt be a calmer, saner, more peaceful sort of place. As Coolidge is supposed to have said, "Don't you know that four-fifths of all our troubles in this life would disappear if we would just sit down and keep still?" (He is also supposed to have said, "If you don't say anything, you won't be called on to repeat it." The only thing a true introvert dislikes more than talking about himself is repeating himself.)

With their endless appetite for talk and attention, extroverts also dominate social life, so they tend to set expectations. In our extrovertist society, being outgoing is considered normal and therefore desirable, a mark of happiness, confidence, leadership. Extroverts are seen as bighearted, vibrant, warm, empathic. "People person" is a compliment. Introverts are described with words like "guarded," "loner," "reserved," "taciturn," "self-contained," "private" — narrow, ungenerous words, words that suggest emotional parsimony and smallness of personality. Female introverts, I suspect, must suffer especially. In certain circles, particularly in the Midwest, a man can still sometimes get away with being what they used to call a strong and silent type; introverted women, lacking that alternative, are even more likely than men to be perceived as timid, withdrawn, haughty.

*Are introverts arrogant?* Hardly. I suppose this common misconception has to do with our being more intelligent, more reflective, more independent, more level-headed, more refined, and more sensitive than extroverts. Also, it is probably due to our lack of small talk, a lack that extroverts often mistake for disdain. We tend to think before talking, whereas extroverts tend to think *by* talking, which is why their meetings never last less than six hours. "Introverts," writes a perceptive fellow named Thomas P. Crouser, in an online review of a recent book called *Why Should Extroverts Make All the Money?* (I'm not making *that* up, either), "are driven to distraction by the semi-internal dialogue extroverts tend to conduct. Introverts don't outwardly complain, instead roll their eyes and silently curse the darkness." Just so.

The worst of it is that extroverts have no idea of the torment they put us through. Sometimes, as we gasp for air amid the fog of their 98-percent-content-free talk, we wonder if extroverts even bother to listen to themselves. Still, we endure stoically, because the etiquette books — written, no doubt, by extroverts — regard declining to banter as rude and gaps in conversation as awkward. We can only dream that someday, when our condition is more widely understood, when perhaps an Introverts' Rights movement has blossomed and borne fruit, it will not be impolite to say "I'm an introvert. You are a wonderful person and I like you. But now please shush."

*How can I let the introvert in my life know that I support him and respect his choice?* First, recognize that it's not a choice. It's not a lifestyle. It's an *orientation.* Second, when you see an introvert lost in thought, don't say "What's the matter?" or "Are you all right?" Third, don't say anything else, either.

CHET RAYMO

# All the Old Sciences Have
# Starring Roles

FROM *The Boston Globe*

WHEN I WAS in high school many long years ago, the sciences were the basics — physics, chemistry, biology. Boys took physics (and went on to become engineers and automobile mechanics), girls took biology (and became nurses and homemakers), and nobody took chemistry if they could help it (except a few nerds who wanted to make stink bombs).

These days the sciences are rather more jumbled up, and students might encounter physical chemistry, biophysics, biochemistry, or any of many other blended specialties. Gender in the science classrooms is rather more jumbled, too.

But, by and large, the old categories stand: If you are going to organize the sciences under a few practical headings, physics, chemistry, and biology are the best way to do it.

These categories are not arbitrary. In broad outline, they correspond to how the universe evolves in space and time.

The universe began as pure physics, in an explosion from an infinitely hot seed of radiant energy. During the first trillion-trillion-trillionth of a second, matter and antimatter flickered in and out of existence. The fate of the universe hung precariously in the balance; it might grow, or it might collapse back into nothingness.

Suddenly it ballooned to enormous size, in what cosmologists call the inflationary epoch, bringing the first particles of matter — the quarks — into existence. Within a millionth of a second, the rapid swelling ceased, and the quarks began to combine into protons and neutrons.

The universe continued to expand and cool, but now at a more stately pace. Within a few more minutes, protons and neutrons combined to form the first atomic nuclei — hydrogen and helium — but the universe was still too hot for the nuclei to attract electrons and make atoms. Not until 300,000 years after the beginning did the first atoms appear.

In all of this, there was nothing of relevance to a chemist or biologist. Chemistry could only begin when parts of the universe had cooled sufficiently for atoms to cling together as molecules (but not so cold that everything is locked together in rock-hard solids). And biology could only begin when the right kind of atoms — carbon, nitrogen, oxygen, phosphorus, and sulfur — had appeared on the scene.

The atoms of life were cooked up in stars by nuclear fusion and blasted into space when the stars died as supernovas. So, even after the era of chemistry had begun, there were not yet the right elements for biological molecules. Billions of years had to pass, and generations of stars had to come and go, before life became a possibility.

Chemistry and biology require a flow of energy, from the fabulously hot interiors of stars to the unimaginable cold of intergalactic space. Only in tiny enclaves precisely placed in this flow is biology possible. The Earth is one such enclave.

Physicists have all the universe to play with. Chemists are basically confined to the galactic neighborhoods of stars. The realm of biologists is those slivers of space — just so far from a star and not too far — where molecules such as water can exist as liquids, somewhere between steam and ice.

If, as most cosmologists now believe, the universe will expand forever, then the sciences will leave the stage in the reverse order in which they made their entrance. As the universe is stretched increasingly thin, the stars will eventually cease to shine and the flow of energy will stop. Life will be extinguished first, then chemical activity.

Physics is indifferent to the temperature of the universe. The infinite temperature of the first instant, or the absolute zero of the end — it is all part of the territory.

High-school science might be a lot more fun if it were taught in the context of the universe's story, The creation myths of our ancestors begin with biology: A humanlike divinity says, "Let there be light." And they end with biology, too; the lights go out with human history. The modern universe story starts and ends with physics, with biology confined somewhere to the middle, and with even chemistry — boring old chemistry — getting its star turn on the stage.

RON ROSENBAUM

# Sex Week at Yale

FROM *The Atlantic Monthly*

CALL ME IS-MALE. You've probably heard about my distant ancestor Ishmael, and the way that when he felt the chill of a "damp, drizzly November" in his soul, he took to following coffins in funeral processions. When Is-male feels a chill in his romantic life, he does something similarly funereal: He tries to give the whole thing the long good-bye. After all, Is-male reasons, he's had more than his fair share of exhilarating encounters — a marriage here, a divorce there — and more than his fair share of misery and heartbreak as well. Why not just draw a line under it, call it done? Style himself after the romantically embittered and disillusioned Graham Greene heroes who trudge off to leper colonies to lose their worldly desires? But Is-male has tried variants of this before, and there has always been some backsliding. He's been looking for something that will put the final nail in the coffin. And then, as if in answer to a prayer, he heard about Sex Week at Yale.

There's nothing like the prospect of a week of academic theorizing about sex and love to make you want to give it all up. And that's *exactly* what I was hoping for when I heard about Sex Week at Yale — lots of theory, lots of abstraction, lots of intellectual distance.

I heard about Sex Week last year in the following press release, forwarded to me by e-mail:

> I'm coordinating a huge event for Yale University which is titled "Campus-Wide Sex Week." Four organizations are organizing the event: Yale Hillel, Peer Health Educators, the Women's Center, and the Yale Lesbian, Gay, Bisexual, Transexual Cooperative . . .
> The week involves a faculty lecture series with topics such as trans-

gender issues: where does one gender end and the other begin, the history of romance, and the history of the vibrator. Student talks on the secrets of great sex, hooking up, and how to be a better lover and a student panel on abstainance. A Valentine's Dinner at the Jewish Center with an afro/cuban band and a debate after the dinner between Rabbi Shmuley Boteach (author of Kosher Sex) and Dr. Judy Kuriansky (radio show host of [Love Phones] and author of [The Complete Idiot's Guide to Tantric Sex]). A faculty panel on sex in college with four professors. a movie film festival (sex fest 2002) and a concert with local bands and yale bands. and lastly, a celebrity panel with Al Goldstein (screw magazine), Dr. Gilda Carle (sex therapist), Nancy Slotnick (Harvard graduate and owner/operator of the Drip Cafe in NYC), and lastly Dr. Susan Block [also a sex-therapist radio host, and a Yale graduate].

The event is going to be huge and all of campus is going to be involved . . .

One of my first thoughts on reading this was that before Yale (my beloved alma mater) had a Sex Week it ought to institute a gala Grammar and Spelling Week. In addition to "abstainance" (unless it was a deliberate mistake in order to imply that "Yale puts the stain in abstinence") there was that intriguing "faculty panel on sex in college with four professors," whose syntax makes it sound more illicit than it was probably intended to be.

But the academic lectures on sex and romance seemed to promise *just* the aversion therapy I was seeking. I know, from keeping up with trendy literary theory, that the more ostensibly "sexual" most academic theorists get, the further from actual sex they get. Thus the disciples of the Parisian postmodern post-Freudian theorist Jacques Lacan are always going on about "desire" and "the body," but one never feels the remotest presence of desire or the body in their thinking. Listening to academics going on about desire is a profound anti-aphrodisiac treasure for those of us seeking surcease from worldly temptations.

## Day One: The Medicalization of Sex

My train arrived in New Haven an hour or so before the first scheduled lecture (on "intersex" issues), which gave me a chance to take a nostalgic sex-scandal tour of the Yale campus and wonder, What the hell is it with Yale and sex?

I first passed the Tomb of Skull and Bones, the legendary secret society that made a ritual of sexual confessions — when not ruling the world, planning the Kennedy assassination, and the like. Seriously, as I once reported, the distinctive element of the Skull and Bones bonding ritual (which two presidents named Bush took part in) is the sexual confessional session, in which each of the initiates must devote one evening to sharing with the other fourteen initiates a detailed history of his sex life. Whether or not they once lay naked in a coffin while talking about sex has not been definitively established. But nudity does figure in another remarkable Yale scandal, one in which I was both exposed and exposer, so to speak, which took place a few blocks north of Skull and Bones, at the Payne Whitney Gymnasium.

This was "The Great Ivy League Nude Posture-Photo Scandal." Yale was not alone in being victimized by the posture-photo scandal: Just about every Ivy League and Seven Sisters school from the 1930s to the 1960s was inveigled into allowing photos of nude or lingerie-clad freshmen to be taken and then transferred to the "research archives" of a megalomaniac pseudo-scientist, W. H. Sheldon. Sheldon believed that the secret of all human character and fate could be reduced to a three-digit number derived from various "postural relationships" (the photos were taken with metal pins affixed to the spine to define the arc of curvature). I was the reporter who discovered, in 1995, that all these nude photos of America's elite — tens of thousands of them, anyway — were available for viewing by "qualified researchers" in an obscure archive of the Smithsonian Institution.

I don't know if this can be classified as a sex scandal, exactly, but it demonstrates the tendency of a certain strain of academic to find a way to abstract from an actual body to a body of mathematical relationships — to pure number rather than impure flesh, if possible.

But to guard against bias during Sex Week at Yale, I made a point of acquiring, confidentially, lecture notes for Harvard's core-curriculum sex course, in order to compare what two of the nation's preeminent schools teach their students about the birds and the bees.

The official title of Harvard's sex course is "Science B-29: Evolution of Human Nature." But according to an impeccable source (a recent graduate), "Everyone just called it 'Sex,' and people would make little jokes to the effect of 'I have Sex right before lunch on Tuesdays and Thursdays.'" That Crimson sense of humor!

Actually, that's unfair. Thumbing through the Harvard course notes on the train to New Haven, I came upon a number of funny, flip remarks by the notetaker about the lecturer's solemn pronouncements on the roots of all human sexual behavior. It seems that Harvard's sex course takes a very strict sociobiological, selfish-gene, evolutionary-psychological, chimp-focused view of human nature — one in which culture and nurture take a back seat to genetics, to millions of years of ingrained *primate* nature. The main variations admitted to exist in human sexual behavior, the chief alternate paths, are based on chimp models. There are the very bad chimpanzees, with their patriarchal "demonic males," and the very good "gentle apes" — the bonobos, pygmy chimps whose females rule. The bonobos have a lot of recreational sex, whereas violence prevails among the demonic males of certain larger primate species. As the Harvard notetaker put it, "So basically Bonobos are a species of female dominated sex freaks. Cool."

The *New York Times* columnist Maureen Dowd once put forward the lovable, sex-positive, violence-averse bonobos as her role models for contemporary society. Interesting: As it turned out, not one of the academic lecturers at Yale's Sex Week had *anything* to say about primates or selfish genes, although on the last day — during the "celebrity panel" — Susan Block, the California sex-advice celebrity, did bring up the bonobos. But in general it could be said that Harvard and Yale were acting out the old nature-versus-nurture split on the nature and nurture of sexuality. For Harvard, your gender behaviors were ruled by your chimp genes. Yale downplayed genetic determinism in favor of cultural contingency: Gender identity was up to *you*.

This was certainly the theme of the first academic lecture of Sex Week. At the designated time, I was the only person in attendance (aside from the lecturer) in the designated room, in Yale's mock-Gothic William L. Harkness Hall. Perhaps it was the title — "The Anatomies of Sex: Theme and Variation" — and the advertised focus on "intersex issues" and hermaphroditism that kept the lecture from being big box office. Suddenly I wondered if Sex Week itself had been a hoax. But then a second person arrived: a fresh-faced young woman bearing a heavily laden backpack, which she set down with relief on the armrest of a chair in the row ahead of me. It soon became apparent what was bulking it up.

"I remembered to bring the sixty-dollar vibrator," she told me, after introducing herself as one of the "peer health educators" — undergraduates who would be leading a seminar and workshop titled "Secrets of Great Sex" for fellow undergraduates after this lecture. She gestured at the vibrator — an anatomically shaped pink latex contraption whose packaging identified it as "The Wascally Wabbit" and whose catchy name appeared to derive from a kind of bunny-ear-shaped attachment.

At this point the lecturer, Professor William Summers, who had been fiddling with his projector apparatus at the front of the classroom, wandered back to greet the peer health educator. Apparently he had attended one of her sex-enhancement seminars, because he spoke enthusiastically about how much he had learned from it.

What I gathered from Professor Summers's lecture was that he is of the school that emphasizes deconstructing the "binary opposition" of male versus female and demonstrating that gender is not some immutable "essence" but a continuum — physiologically, culturally, and psychologically. For the Yale continuum school, the spotlight is on the interzone of "intersex" people, as well as transsexuals and transvestites, because they demonstrate that the binary view of gender is false. Of course, you could also say that the exceptions prove the rule — the intersexual anomalies remind us of the binary poles — but at Yale they don't: The exceptions *are* the rule.

Enter the hermaphrodites. One *could* say that the sexual theorists of the intersex school, with their close focus on the genitals of hermaphrodites (or, as some prefer to be called, intersexuals), are allowing the tail to wag the dog. In the past the orthodox wisdom about those who were born hermaphrodites was that they should be surgically "corrected" and assigned a single sexual identity. Now, it seems, hermaphrodites are demanding that they be viewed as normal, just another point on the continuum.

Anyway, such was the theme of *Hermaphrodites Speak!*, the first documentary about hermaphrodites made *by* hermaphrodites, "filmed at an intersex retreat" — a film shown by Professor Summers after he treated us to huge blowups of hermaphroditic genitals at all stages of growth. Much of the film consisted of a half dozen or so hermaphrodites sitting around a picnic blanket in a rustic setting sharing their feelings, describing the struggle for self-esteem and their not-so-residual anger at the surgeons who tried to use a scalpel to "correct" them into a single sex. One was talking in bitter tones about his/her desire to take "a dull rusty knife" to the genitals of his/her surgeon. It was at about this time that the "Jazz on Film" contretemps developed.

This is how it happened: Professor Summers hadn't been able to get his PowerPoint software — with which he was to project images from his laptop onto the lecture-hall screen — to work in the originally designated lecture room. Tech help from the crack Yale audio-visual squad had been summoned, and it was determined that the problem was not in Professor Summers's equipment but in the tech infrastructure of that lecture room. A mass shift was made to another classroom. Well, "mass shift" is a bit of an exaggeration: You could still count the number who had finally showed up on the fingers of one hand.

Everything was proceeding smoothly in the new lecture hall, which had been empty at the time of the move, until a half hour or so into the intersex presentation, when students began drifting in expecting a previously scheduled seminar in that room: "Jazz on Film." I could tell as they came in and took seats that they were trying to reconcile what they expected to see onscreen (Mingus and Monk, maybe, talking about the history of bebop) with what they were actually seeing — slides of hermaphroditic genitals, including "ovo-testes," and then *Hermaphrodites Speak!* There was a bit of cognitive dissonance to be resolved. I suppose that, in a way, the whole intersex/hermaphrodite phenomenon could be considered Nature's jazz — not defect but improvisation.

In any case, after some of the "Jazz on Film" students began to realize that this was, well, something else again, an agreement was reached that the intersex lecture would move on, like a gypsy, to yet another classroom, for those who wanted not to miss the last half of *Hermaphrodites Speak!*

I must admit I gave it a pass: I had just enough time to catch the next train back to New York, satisfied that the first day of Sex Week at Yale had exceeded my aversion-therapy expectations.

## *Day Two: The Materialist Anthropology of Love and Sex*

The second faculty lecture of Sex Week was titled "The History and Theory of Romantic Love in European Culture." At first it seemed to me that Professor Linda-Anne Rebhun's perspective was more down-to-earth than that of Professor Summers, whose lecture was shot through with academic ideology. Down-to-earth even if it was a remote part of the earth. Rebhun is an anthropologist specializing in the effects of urbanization on those folk in Brazil who have left traditional life in rural villages for the cities and are adjusting their notions of sex, romance, and marriage to their new environment.

Although her Brazilian field notes might not seem at first to have a direct bearing on "The History and Theory of Romantic Love in European Culture," the implicit thesis was that something akin to the sudden adoption of the notion of romantic love, which occurred centuries ago in European culture, could be glimpsed going on right now in real time with a set of newly urbanized, newly middle-class types in Brazil. It fit in with a currently fashionable academic mode of thinking that I recently saw referred to as "inventionism."

Inventionism has been rife in the study of romance ever since C. S. Lewis declared in *The Allegory of Love* that romantic love *did not exist* until it was "invented" by the poets known as troubadours in eleventh-century Provence, who created the rituals and rules of courtly love and embodied them in their poems; it was, in effect, the poems that *created* love. Love was a literary convention that became an emotion. I've never understood how a scholar of Lewis's erudition could believe such nonsense — particularly a scholar who had read Catullus's tormented love lyrics, written a thousand years before the Provençal troubadours, or Ovid's *Amores,* or Virgil's Aeneas and Dido episode. Still, inventionism rules in the humanities, the obvious appeal being that it can allow scholars to become inventors themselves.

I was thrown off a bit by Professor Rebhun. She was wearing a lovely but discreet silver heart-shaped pin that suggested a belief in love. And yet she was asserting that love was merely an illusion imposed by the social structure: that the indigenous Brazilians didn't have love as we know it until economic pressures made it *functional.* The need to "commodify sexual desire" in an urban context — to embed childbearing in a capitalistic system of ownership and property transfer — required the invention of a fiction called love.

What I took from the lecture was that it's really all about money, if you look at it in a hardheaded cultural-materialist, anthropological way. One could quarrel with the reductive reasoning of this materialist-inventionist view of love and sex. One could ask just how "economic pressures" — these cold abstractions — could possibly invent (as opposed to influence) such profound and complex states as romantic and erotic love. Does it all come down to "We are living in a material world"? And we are all material girls, so to speak?

The professor's lecture made me think of another song, one my parents sang together when we were on a drive or sometimes when they were just standing around the kitchen. It was their favorite song, the one I remember as defining the love that kept them together for forty-five years, till death did them part: "Side by Side."

> Oh, we ain't got a barrel of money.
> Maybe we're ragged and funny,
> But we'll travel along
> Singin' a song,
> Side by side.

Hearing them sing that refrain made me believe in love even before I met my first supermodel. (Just kidding.) But I grew up believing that love was more than property relations, and not a delusional product of them.

The professor continued her grim disquisition on the specious, epiphenomenal nature of romantic love. Not only was it a delusion, she said, but it could be a *deadly* delusion, because so many of the great romantic loves in literature, from Tristan and Isolde to Romeo and Juliet, always and inevitably culminated in, aimed toward — were virtually fulfilled in — "mutual suicide." Love — the suicide bomb of the emotions. There is an inevitable disconnect, she told us, between our idea of romance and the stable relationship of marriage. Abandon all hope of love, ye who enter into wedlock. Our ideas of romantic love are incongruent not just with marriage but with *all* long-term relationships, not to mention with child-rearing. The whole concept of romance can be seen as "an attempt to beautify lust."

Her pessimism, her "realism," her reductionism, her materialism, were so relentless that even though I was seeking to hear words that would help to *extinguish* the last embers of longing, I felt those embers flare up in rebellion. This lecture was a total multicultural, multidisciplinary, multiphasic attack on the reality of romantic love. And I wish it had been more convincing, because what I remember most from this lecture was the motion-detector problem, which became for me a kind of metaphor.

There were technical difficulties with this room, too, though they were of a nature different from the ones Professor Summers had encountered. The lights kept going out at what seemed like significant moments in Professor Rebhun's lecture. She'd be talking about Tristan and Isolde and mutual suicide, and suddenly the room would go dark. It turned out that this was an attendance problem. There were so few in the lecture room to hear the deconstruction of romance that the sensors that were designed to keep the lights on as long as they detected some motion — some sign of life — would periodically shut the lights off because they detected none. I suspect that if the room had instead featured an "emotion detector," Professor Rebhun's materialism would have turned out the lights for good.

## Day Three: The History of the Vibrator

Call me a coward, but I skipped Day Three. As I understand it, the lecture was a serious academic discourse on the evolution of vibrator technology, from post–Civil War steam-powered contraptions (no joke!) to today's superefficient Wascally Wabbits, Wile E. Coyotes, and Elmer Fudds (note to fact-checking: I made up the last two).

But I did begin to see a pattern developing in the academic lectures. Day One: the medicalization of sex. Day Two: the materialization (or monetization) of sex. Day Three: the mechanization of sex. And then . . .

## Day Four: The Spiritualization of Sex

Thursday, Valentine's Day, was supposed to end in a big debate, and this time on the train to New Haven my Sex Week homework included heavyweight, embarrassing-to-read-on-the-train tomes by the featured debaters: *The Complete Idiot's Guide to Tantric Sex,* by Judy Kuriansky, who is described in that book as an adjunct professor at Columbia and a sex therapist; and *Kosher Sex,* by Rabbi

Shmuley Boteach, the former spiritual adviser to Oxford's Jewish community and a best-selling sex-and-love-advice author.

Dr. Judy's book was a curious combination of the spiritual and the technical, and even though I'd resolved to leave the pleasures of the flesh behind, I am always open to learning new things. I have to admit I came across two technical terms in *The Complete Idiot's Guide* that I had not encountered before: the "Kivin method" and the "X spot."

I think I'm just *not* going to get into the Kivin method in this story; I'm trying for an R rating. As for the X spot, it, too, is X-rated, but at least I can speak of it by analogy with the G spot (which isn't exactly G-rated). The X spot, Dr. Judy tells us, is like the G spot but in another location — an important location that I won't reveal to you (and I'm keeping quiet about the "AFE zone" and the "PFE zone," too). But this brave new venture into alphabetical terra incognita reminded me of one of the more unusual reporting experiences I've had, doing a story I never wrote up: a story about the sixth World Congress of Sexology, which was held in Washington, D.C., in the mid 1980s, just about the time when the G spot was coming into prominence and people were still debating whether it existed and what it was for.

"Sexology": It's too bad it's such a silly-sounding moniker, one that makes the whole profession seem like a Monty Python sketch. And yet the scientific study of sex (as opposed to unscientific, Kinsey-type sexology) has a place in a world that for centuries lived in ignorance about the way things work "down there" (to use the technical term). Although it may be true that we suffer from too much talk about sex, we don't necessarily have too much information.

What I remember about the World Congress of Sexology is that the debate over the existence of the G spot was similar to the argument of medieval cartographers over the location of the western edge of the world. I remember reeling from all the scientific sex talk and then going almost immediately to one of my Yale class reunions, where I mentioned the G-spot debate. The wife of one of my classmates declared forthrightly, "I don't know what the debate's about; *I have one!*"

And now, according to Dr. Judy, we have the X spot — if not a new place on the map, a new *name* for a place whose existence anecdotal evidence, let us say, has suggested. Of course, there's an argument for discovering these alphabet spots in innocence and wonder, although in reality innocence and wonder can be a euphemism for ignorance and error. (But I can't wait to ask about the X spot at my next Yale reunion.)

So I applaud the investigative-sexologist side of Dr. Judy, but that was not the side on display during her appearance at the Yale debate. The Valentine's Night debate — tantric versus kosher sex — took place at the Joseph Slifka Center for Jewish Life at Yale. Dr. Judy entered wearing an electric-blue blazer and turquoise jewelry, and she radiated media-celeb empowerment. (She had to interrupt her talk to do a live feed to CNN, she told us.) Instead of giving a conventional debate presentation, she put on a tape of Bette Midler singing "The Rose," and a selection of Lilith Fair favorites, and led the fifty or so attendees through a series of touchy-feely "intimacy exercises," lining up strangers to hold their hands on one another's hearts and stare into one another's eyes and perform other California hot-tub maneuvers. This seemed to drive Rabbi Boteach out of the room, although he later explained that he'd had a pressing need to phone his family.

What was curious about Dr. Judy's talk, interspersed with the intimacy exercises, was her insistence that tantric sex isn't really about sex, or anything, you know, dirty like that — as if it would be somehow shameful if it were about sex, or certainly something lesser than the religious experience that tantric practices turn sex into, thus sanctifying it.

"Tantric sex is . . . not about the penis and vagina," Dr. Judy declared. "It's using sex to move energy, to activate higher states of bliss." Sex, even tantric sex, is just a tool, so to speak, to lift us above sex; the sexual energy is a "generator" — which somehow, inappropriately I'm sure, made me think of those steam-powered vibrators. (It also made me think of the Woody Allen movie where his character is asked by a psychiatrist, "Do you think sex is dirty?" and he replies, "It is if you're doing it right.")

One got the inescapable impression from Dr. Judy that sex was good only when it was goody-goody, spiritualized. Rabbi Boteach was not much different, in his own kosher way. You know about Rabbi Boteach, right? He's the guy who started out as an earnest young Orthodox rabbinical adviser to Oxford students and then somehow, by a process that is still a bit mysterious to me (involving, at a crucial point, Michael Jackson's visit to Oxford), morphed into that special contemporary phenomenon, the "celebrity friend" — of Michael and Uri Geller and the whole Hollywood kabbalah crowd. All of which he parlayed into best-selling books, incessant TV-talk-show appearances, and more best sellers, until . . . Rabbi Boteach confessed to the world that he felt used and burned out, and decided that he must step off the celebrity carousel to regain his balance. It was this new, subdued Rabbi Boteach that we were seeing. Although he still had names to drop, he made a point of dropping the name of Sir Isaiah Berlin ("When I was at Oxford, I asked Sir Isaiah . . .") before dropping the name of Madonna.

Still, Rabbi Boteach came across as more intelligent and self-deprecating than his previous relentless self-promotion suggested. He displayed an awareness that the realities of making marriage work involved more than kabbalah and sanctification. He even invoked "webcam girls," from reality-porn sites, in the context of what he said would be his next book: *Kosher Adultery*. The concept here seemed to be to make marriage and monogamy "dirty" (in a good way) as a means of addressing a problem he had evidently heard about from those seeking his counsel: that some husbands can be more lustfully aroused by "webcam girls" than by their wives. "Kosher adultery," I suspected, involves what has often been called "role playing" (or as Rabbi Boteach put it, "bringing the erotic energy of unfaithfulness into marriage"), getting the wives to act like "other women." But when *Kosher Adultery* came out, later in the year, I was surprised to see it emphasize the notion that *husbands* should fantasize that their wives were with other *men*. So the wives are playing other women, the husbands are playing other men . . . it takes a lot of work to preserve Shmuleyan monogamy.

Still, I admired Rabbi Boteach's willingness to explore these complex questions. But then I was disappointed when he proceeded to follow a path similar to Dr. Judy's: He abstracted from real people with real bodies to mystical circles and lines. In a way his approach is not unlike Harvard's chimpification of sex, at least in degree of abstraction, although instead of appealing to science it creates a supernatural, kabbalistic gloss — something to do with the fact that God starts out as a circle, an infinity that can't be measured or apprehended. But to create finite creatures "God had to withhold, to vacate Himself to radiate a finite light, a line that represents the masculine aspect of God, Elohim." The Shekinah, or visible manifestation of God, in kabbalah is the feminine aspect, God as circle. Because "men are lines and women are circles," their sexual natures reflect that. "Men are linear, women are cyclical."

Talk about your reflexive essentialism. "The number ten is made up of a circle and a line, which makes it a vision of perfection," Rabbi Boteach averred, sounding like Louis Farrakhan discoursing on the number nineteen. The world is supposed to be about the supremacy of the feminine, "because love is greater than justice," Rabbi Boteach said, now descending into John Gray-style Mars-and-Venus psychobabble.

It was then, after all this talk of sacred lines and circles, that I thought of Ronald Reagan. Not in *that* way. Instead I remembered a certain slogan that became popular among "movement conservatives" in the eighties — "Let Reagan be Reagan!" And after this double dose of tantric and kosher sanctification of sex following on mechanization, materialization, medicalization, and chimpification, it made me want to say, in effect, "Let Reagan be Reagan": *Let sex be sex.*

Why won't the experts let sex be sex? An answer to that question suggested itself on . . .

## Day Five: The Day of "Sex with Four Professors"

This turned out, to my surprise, to be one of the least well attended but most illuminating events of Sex Week. When I say it wasn't well attended, let me put it this way: The attendees, if you exclude the two earnest and thoughtful undergraduate organizers of Sex Week (Eric Rubenstein and Jacqueline Farber), were outnumbered by the four panelists.

Seeing the utter emptiness of the venue, the four panelists, in consultation with the two organizers, decided to turn the event into a conversation among themselves. And it wasn't a bad conversation at all, because the four professors represented, for the most part, the humanistic tradition in Yale social-science thinking about sex. One could hear echoes of Erik Erikson's "identity psychology" as one of the professors spoke of the difficulty college students found in suddenly adjusting their "affectional currents" —

directed so long to family members — to sexual partners. (This seemed a little unrealistic in an age when suburban kids are playing oral-sex versions of Spin the Bottle at age fourteen, although that may not be the kind of affectional current the professor was referring to.)

There was also much optimistic talk about the ways in which "Norwegian studies" have shown that "beginning sex education as early as kindergarten" is a Good Thing. All things Scandinavian are Good Things to sex experts: so wholesome, so rational.

But something more interesting emerged from the "sex with four professors" panel. Perhaps it was because one of the professors was a psychiatric counselor and another was a minister. These people had been in the trenches dealing with troubled students. They had not dealt only with the optimistic sex-as-enlightenment, sex-as-sanctifying-gospel, sex-as-Norwegian-kindergarten view of things; they had come face-to-face with young people who had been badly hurt, emotionally devastated, by sexual experiences in college.

I forget which one of them said it, but in my notes is the line "Turned upside down sex can be a *nightmare*. Sex can have *terrible* effects." There is another line in the same vein: "Jealousy is the most powerful emotion." That's a major statement: Jealousy, the misery-inducing derivative of love and lust, is more powerful than its progenitors. The tail wagging the dog again.

Here, perhaps, is the explanation for why nobody really wants to let Reagan be Reagan when it comes to sex: because it *can* be dangerous, even devastating, not a blissful "energy transfer" — and if an "energy transfer" at all, then more like stepping on the third rail.

## Day Six

I skipped "Sex Fest 2002," the film festival. Instead I went to the video store and rented the film I recalled as being at once the most erotic and the bleakest and darkest vision of sex on celluloid, one whose very title sums up the destructive third-rail power of sexuality: *Damage*. Jeremy Irons and Juliette Binoche destroy themselves and their families in the grip of an insane but utterly convincing passion (she's his son's fiancée!). Just the kind of thing Norwegian kindergarten fails to prepare you for.

*Day Seven: The "Celebrity Panel"*

I'm not sure you could call them celebrities outside the sex, love, and pornography realm. But I was glad I attended, because there were several surprising revelations from the celebrity panel — and one that was truly shocking.

Dr. Susan Block, dressed to thrill with a bare midriff and a cowboy hat, told the audience (more than the usual handful) that she was a Yale graduate and that "we didn't have Sex Week when I was at Yale — we had sex all year round." She also boasted, "We had sex with teachers, and it was a good way of learning more from them."

Dr. Block spoke up for the old values of the sixties sexual revolution: "Sex is revolutionary!" she told the audience. "Nietzsche said that the spiritualization of sex is love." Then she shifted into a paean to Harvard's favorite primates, the bonobos, the feminist heroines of the chimp world — "extremely sexual" primates, she told us, "who have learned to use sex to defuse violence."

Nancy Slotnick, a Harvard graduate who invented a café and dating service on Manhattan's Upper West Side, called Drip, spoke up for love, albeit in a curious way: "Love is women's favorite fetish," she said. "A fetish is something you don't need to get sex but you feel you need to enjoy it."

But it was Al Goldstein who provided a fitting climax to Sex Week. Goldstein at first seemed to take pleasure in bringing his obscenity-riddled view of sex to Yale's august precincts, making indecent proposals to the women panelists while boasting that his son was "a Harvard man, who's graduating from Harvard Law School this weekend at the top of his class." He talked a good game, all the while keeping in character (obscene wise-guy loudmouth), but the subtext suggested that he was just an old softie. Well, maybe that's not the most apt phrase to use. How about "suggested that he was a romantic at heart"? An embittered romantic, but a romantic.

Goldstein disclosed an astonishing autobiographical fact: The publisher of *Screw*, the self-proclaimed shameless pornographer, has been married *four times!* Now, they say that a second marriage is "the triumph of hope over experience." But third and fourth marriages are more generally the triumph of addiction over everything else. Addiction to love. Why four marriages *unless* for love?

It may be the ultimate tribute to the power of love — perhaps the most powerful tribute since *Antony and Cleopatra,* a tale about two other much married people possessed by addiction to love. It suggests that love triumphs over the lure of pornography; love is more seductive than webcam girls, faster than a speeding bullet, more powerful than a locomotive — not the suicide bomb but the superhero of emotions. Of course, Al Goldstein would never come out and say it. "Men are doomed!" he shouted to the bewildered Yale audience. "Love is *evil.*" He was like the maddened King Lear of lust, raging against his most dangerous foe: "Do thy worst, blind Cupid," Lear declares. "I'll not love."

I'm not sure whether Goldstein really meant it or whether it was the bitter love-hate cry of the hopeless addict for his one true fix. Still, it was just what I wanted to hear. It was music to my ears. Love is *evil.* Thank you, Al.

STEVE SAILER

# The Cousin Marriage Conundrum

FROM *The American Conservative*

MANY PROMINENT neoconservatives are calling on America not only to conquer Iraq (and perhaps more Muslim nations after that), but also to rebuild Iraqi society in order to jumpstart the democratization of the Middle East. Yet, Americans know so little about the Middle East that few of us are even aware of one of the building blocks of Arab Muslim cultures — cousin marriage. Not surprisingly, we are almost utterly innocent of any understanding of how much the high degree of inbreeding in Iraq could interfere with our nation building ambitions.

In Iraq, as in much of the region, nearly half of all married couples are first or second cousins to each other. A 1986 study of 4,500 married hospital patients and staff in Baghdad found that 46 percent were wed to a first or second cousin, while a smaller 1989 survey found 53 percent were "consanguineously" married. The most prominent example of an Iraqi first cousin marriage is that of Saddam Hussein and his first wife Sajida.

By fostering intense family loyalties and strong nepotistic urges, inbreeding makes the development of civil society more difficult. Many Americans have heard by now that Iraq is composed of three ethnic groups — the Kurds of the north, the Sunnis of the center, and the Shi'ites of the south. Clearly, these ethnic rivalries would complicate the task of ruling or reforming Iraq. But that's just a top-down summary of Iraq's ethnic makeup. Each of those three ethnic groups is divisible into smaller and smaller tribes, clans, and inbred extended families — each with their own alliances, rivals, and feuds. And the engine at the bottom of these bedeviling social divisions is the oft-ignored institution of cousin marriage.

The fractiousness and tribalism of Middle Eastern countries have frequently been remarked. In 1931, King Feisal of Iraq described his subjects as "devoid of any patriotic idea, . . . connected by no common tie, giving ear to evil; prone to anarchy, and perpetually ready to rise against any government whatever." The clannishness, corruption, and coups frequently observed in countries such as Iraq appear to be tied to the high rates of inbreeding.

Muslim countries are usually known for warm, devoted extended family relationships, but also for weak patriotism. In the United States, where individualism is so strong, many assume that "family values" and civic virtues such as sacrificing for the good of society always go together. But, in Islamic countries, family loyalty is often at war with national loyalty. Civic virtues, military effectiveness, and economic performance all suffer.

Commentator Randall Parker wrote:

> Consanguinity [cousin marriage] is the biggest underappreciated factor in Western analyses of Middle Eastern politics. Most Western political theorists seem blind to the importance of pre-ideological kinship-based political bonds in large part because those bonds are not derived from abstract Western ideological models of how societies and political systems should be organized. . . . Extended families that are incredibly tightly bound are really the enemy of civil society because the alliances of family override any consideration of fairness to people in the larger society. Yet, this obvious fact is missing from 99 percent of the discussions about what is wrong with the Middle East. How can we transform Iraq into a modern liberal democracy if every government worker sees a government job as a route to helping out his clan at the expense of other clans?

Retired U.S. Army colonel Norvell De Atkine spent years trying to train America's Arab allies in modern combat techniques. In an article in *American Diplomacy* entitled, "Why Arabs Lose Wars," a frustrated De Atkine explained,

> First, the well-known lack of trust among Arabs for anyone outside their own family adversely affects offensive operations. In a culture in which almost every sphere of human endeavor, including business and social relationships, is based on a family structure, this basic mistrust of others is particularly costly in the stress of battle. Offensive action, at base, consists of fire and maneuver. The maneuver element must be confident that supporting units or arms are providing covering fire. If there is a lack of trust in that support, getting troops moving forward against dug-

in defenders is possible only by officers getting out front and leading, something that has not been a characteristic of Arab leadership.

Similarly, as Francis Fukuyama described in his 1995 book *Trust: The Social Virtues and the Creation of Prosperity,* countries such as Italy with highly loyal extended families can generate dynamic family firms. Yet, their larger corporations tend to be rife with goldbricking, corruption, and nepotism, all because their employees don't trust each other to show their highest loyalty to the firm rather than their own extended families. Arab cultures are more family-focused than even Sicily, and thus their larger economic enterprises suffer even more.

American society is so biased against inbreeding that many Americans have a hard time even conceiving of marrying a cousin. Yet, arranged matches between first cousins (especially between the children of brothers) are considered the ideal throughout much of a broad expanse from North Africa through West Asia and into Pakistan and India.

In contrast, Americans probably disapprove of what scientists call "consanguineous" mating more than any other nationality. Three huge studies in the United States between 1941 and 1981 found that no more than 0.2 percent of all American marriages were between first cousins or second cousins.

Americans have long dismissed cousin marriage as something practiced only among hillbillies. That old stereotype of inbred mountaineers waging decades-long blood feuds had some truth to it. One study of 107 marriages in Beech Creek, Kentucky, in 1942 found 19 percent were consanguineous, although the Kentuckians were more inclined toward second-cousin marriages, while first-cousin couples are more common than second-cousin pairings in the Islamic lands.

Cousin marriage averages not much more than 1 percent in most European countries, and under 10 percent in the rest of the world outside that Morocco-to-southern-India corridor.

Muslim immigration, however, has been boosting Europe's low level of consanguinity. According to the leading authority on inbreeding, geneticist Alan H. Bittles of Edith Cowan University in Perth, Australia, "In the resident Pakistani community of some 0.5 million [in Britain] an estimated 50 percent to 60-plus percent of marriages are consanguineous, with evidence that their prevalence is increasing." (Bittles's Web site, www.consang.net, presents the results of several hundred studies of the prevalence of inbreeding around the world.)

European nations have recently become increasingly hostile toward the common practice among their Muslim immigrants of arranging marriages between their children and citizens of their home country, frequently their relatives. One study of Turkish guest workers in the Danish city of Ishøj found that 98 percent — first, second, and third generation — married a spouse from Turkey who then came and lived in Denmark. (Turks, however, are quite a bit less enthusiastic about cousin marriage than are Arabs or Pakistanis, which correlates with the much stronger degree of patriotism found in Turkey.)

European "family reunification" laws present an immigrant with the opportunity to bring in his nephew by marrying his daughter to him. Not surprisingly, "family reunification" almost always works just in one direction — with the new husband moving from the poor Muslim country to the rich European country.

If a European-born daughter refused to marry her cousin from the old country just because she doesn't love him, that would deprive her extended family of the boon of an immigration visa. So, intense family pressure can fall on the daughter to do as she is told.

The new Danish right-wing government has introduced legislation to crack down on these kinds of marriages arranged to generate visas. British Home Secretary David Blunkett has called for immigrants to arrange more marriages within Britain.

Unlike the Middle East, Europe underwent what Samuel P. Huntington calls the "Romeo and Juliet revolution." Europeans became increasingly sympathetic toward the right of a young woman to marry the man she loves. Setting the stage for this was the Catholic Church's long war against cousin marriage, even out to fourth cousins or higher. This weakened the extended family in Europe, thus lessening the advantages of arranged marriages. It also strengthened the nuclear family as well as broader institutions like the Church and the nation-state.

Islam itself may not be responsible for the high rates of inbreeding in Muslim countries. (Similarly high levels of consanguinity are found among Hindus in southern India, although there, uncle-niece marriages are socially preferred, even though their degree of genetic similarity is twice that of cousin marriages, with worse health consequences for offspring.)

Rafat Hussain, a Pakistani-born senior lecturer at the University of New England in Australia, told me, "Islam does not specifically encourage cousin marriages and, in fact, in the early days of the spread of Islam, marriages outside the clan were highly desirable to increase cultural and religious influence." She adds, "The practice has little do with Islam (or in fact any religion) and had been a prevalent cultural norm before Islam." Inbreeding (or "endogamy") is also common among Christians in the Middle East, although less so than among Muslims.

The Muslim practice is similar to older Middle Eastern norms, such as those outlined in Leviticus in the Old Testament. The lineage of the Hebrew patriarchs who founded the Jewish people was highly inbred. Abraham said his wife Sarah was also his half-sister. His son Isaac married Rebekah, a cousin once removed. And Isaac's son Jacob wed his two first cousins, Leah and Rachel.

Jacob's dozen sons were the famous progenitors of the Twelve Tribes of Israel. Due to inbreeding, Jacob's eight legitimate sons had only six unique great-grandparents instead of the usual eight. That's because the inbred are related to their relatives through multiple paths.

Why do so many people around the world prefer to keep marriage in the family? Hussain noted, "In patriarchal societies where parents exert considerable influence and gender segregation is followed more strictly, marriage choice is limited to whom you know. While there is some pride in staying within the inner bounds of family for social or economic reasons, the more important issue is: Where will parents find a good match? Often, it boils down to whom you know and can trust."

Another important motivation — one that is particularly important in many herding cultures, such as the ancient ones from which the Jews and Muslims emerged — is to prevent inheritable wealth from being split among too many descendants. This can be especially important when there are economies of scale in the family business.

Just as the inbred have fewer unique ancestors than the outbred, they also have fewer unique heirs, helping keep both the inheritance and the brothers together. When a herd-owning patriarch marries his son off to his younger brother's daughter, he insures that his grandson and his grandnephew will be the same person. Likewise, the younger brother benefits from knowing that his grandson will also be the patriarch's grandson and heir. Thus, by making sibling rivalry over inheritance less relevant, cousin marriage emotionally unites families.

The anthropologist Carleton Coon also pointed out that by minimizing the number of relatives a Bedouin Arab nomad has, this system of inbreeding "does not overextend the number of persons whose deaths an honorable man must avenge."

Of course, there are also disadvantages to inbreeding. The best known is medical. Being inbred increases the chance of inheriting genetic syndromes caused by malign recessive genes. Bittles found that, after controlling for socioeconomic factors, the babies of first cousins had about a 30 percent higher chance of dying before their first birthdays.

The biggest disadvantage, however, may be political.

Are Muslims, especially Arabs, so much more loyal to their families than to their nations because, due to countless generations of cousin marriages, they are so much more genealogically related to their families than Westerners are related to theirs? Frank Salter, a political scientist at the Max Planck Institute in Germany whose new book *Risky Transactions: Trust, Kinship, and Ethnicity* takes a sociobiological look at the reason why Mafia families are indeed families, told me, "That's my hunch; at least it's bound to be a factor."

One of the basic laws of modern evolutionary science, quantified by the great Oxford biologist William D. Hamilton in 1964 under the name "kin selection," is that the more close the genetic relationship between two people, the more likely they are to feel loyalty and altruism toward each other. Natural selection has molded us not just to try to propagate our own genes, but to help our relatives, who possess copies of some of our specific genes, to propagate their own.

Nepotism is thus biologically inspired. Hamilton explained that the level of nepotistic feeling generally depends upon degree of genetic similarity. You share half your personally variable genes with your children and siblings, but one quarter with your nephews/

nieces and grandchildren, so your nepotistic urges will tend to be somewhat less toward them. You share one eighth of your genes with your first cousins, and one thirty-second with your second cousin, so your feelings of family loyalty tend to fall off quickly.

But not as quickly if you and your relatives are inbred. Then, you'll be genealogically related to your kin via multiple pathways. You will all be genetically more similar, so your normal family feelings will be multiplied. For example, your son-in-law might also be the nephew you've cherished since his childhood, so you can lavish all the nepotistic altruism on him that in outbred Western societies would be split between your son-in-law and your nephew.

Unfortunately, nepotism is usually a zero-sum game, so the flip side of being materially nicer toward your relatives would be that you'd have less resources left with which to be civil, or even just fair, toward non-kin. So, nepotistic corruption is rampant in countries such as Iraq, where Saddam has appointed members of his extended family from his hometown of Tikrit to many key positions in the national government.

Similarly, a tendency toward inbreeding can turn an extended family into a miniature racial group with its own partially isolated gene pool. (Dog breeders use extreme forms of inbreeding to quickly create new breeds in a handful of generations.) The ancient Hebrews provide a vivid example of a partly inbred extended family (that of Abraham and his brothers) that evolved into its own ethnic group. This process has been going on for thousands of years in the Middle East, which is why not just the Jews but also tiny, ancient inbreeding groups such as the Samaritans and the John the Baptist–worshiping Sabeans still survive.

In summary, although neoconservatives constantly point to America's success at reforming Germany and Japan after World War II as evidence that it would be easy to do the same in the Middle East, the deep social structure of Iraq is the complete opposite of those two true nation-states, with their highly patriotic, cooperative, and (not surprisingly) outbred peoples. The Iraqis, in contrast, more closely resemble the Hatfields and the McCoys.

ROBERT SAPOLSKY

# Bugs in the Brain

FROM *Scientific American*

LIKE MOST SCIENTISTS, I attend professional meetings every now and then, one of them being the annual meeting of the Society for Neuroscience, an organization of most of the earth's brain researchers. This is one of the more intellectually assaulting experiences you can imagine. About 28,000 of us science nerds jam into a single convention center. After a while, this togetherness can make you feel pretty nutty: For an entire week, go into any restaurant, elevator, or bathroom, and the folks standing next to you will be having some animated discussion about squid axons. The process of finding out about the science itself is no easier. The meeting has 14,000 lectures and posters, a completely overwhelming amount of information. Of the subset of those posters that are essential for you to check, a bunch remain inaccessible because of the enthusiastic crowds in front of them, one turns out to be in a language you don't even recognize, and another inevitably reports every experiment you planned to do for the next five years. Amid it all lurks the shared realization that despite zillions of us slaving away at the subject, we still know squat about how the brain works.

My own low point at the conference came one afternoon as I sat on the steps of the convention center, bludgeoned by information and a general sense of ignorance. My eyes focused on a stagnant, murky puddle of water by the curb, and I realized that some microscopic bug festering in there probably knew more about the brain than all of us neuroscientists combined.

My demoralized insight stemmed from a recent extraordinary paper about how certain parasites control the brain of their host.

Most of us know that bacteria, protozoa, and viruses have astonishingly sophisticated ways of using animal bodies for their own purposes. They hijack our cells, our energy, and our lifestyles so they can thrive. But in many ways, the most dazzling and fiendish thing that such parasites have evolved — and the subject that occupied my musings that day — is their ability to change a host's behavior for their own ends. Some textbook examples involve ectoparasites, organisms that colonize the surface of the body. For instance, certain mites of the genus *Antennophorus* ride on the backs of ants and, by stroking an ant's mouthparts, can trigger a reflex that culminates in the ant's disgorging food for the mite to feed on. A species of pinworm of the genus *Syphacia* lays eggs on a rodent's skin, the eggs secrete a substance that causes itchiness, the rodent grooms the itchy spot with its teeth, the eggs get ingested in the process, and once inside the rodent they happily hatch.

These behavioral changes are essentially brought about by annoying a host into acting in a way beneficial to the interlopers. But some parasites actually alter the function of the nervous system itself. Sometimes they achieve this change indirectly, by manipulating hormones that affect the nervous system. There are barnacles *(Sacculina granifera)*, a form of crustacean, found in Australia that attach to male sand crabs and secrete a feminizing hormone that induces maternal behavior. The zombified crabs then migrate out to sea with brooding females and make depressions in the sand ideal for dispersing larvae. The males, naturally, won't be releasing any. But the barnacles will. And if a barnacle infects a female crab, it induces the same behavior — after atrophying the female's ovaries, a practice called parasitic castration.

Bizarre as these cases are, at least the organisms stay outside the brain. Yet a few do manage to get inside. These are microscopic ones, mostly viruses rather than relatively gargantuan creatures like mites, pinworms, and barnacles. Once one of these tiny parasites is inside the brain, it remains fairly sheltered from immune attack, and it can go to work diverting neural machinery to its own advantage.

The rabies virus is one such parasite. Although the actions of this virus have been recognized for centuries, no one I know of has framed them in the neurobiological manner I'm about to. There are lots of ways rabies could have evolved to move between hosts.

The virus didn't have to go anywhere near the brain. It could have devised a trick similar to the one employed by the agents that cause nose colds — namely, to irritate nasal-passage nerve endings, causing the host to sneeze and spritz viral replicates all over, say, the person sitting in front of him or her at the movies. Or the virus could have induced an insatiable desire to lick someone or some animal, thereby passing on virus shed into the saliva. Instead, as we all know, rabies can cause its host to become aggressive so the virus can jump into another host via saliva that gets into the wounds.

Just think about this. Scads of neurobiologists study the neural basis of aggression: the pathways of the brain that are involved, the relevant neurotransmitters, the interactions between genes and environment, modulation by hormones, and so on. Aggression has spawned conferences, doctoral theses, petty academic squabbles, nasty tenure disputes, the works. Yet all along, the rabies virus has "known" just which neurons to infect to make a victim rabid. And as far as I am aware, no neuroscientist has studied rabies specifically to understand the neurobiology of aggression.

Despite how impressive these viral effects are, there is still room for improvement. That is because of the parasite's nonspecificity. If you are a rabid animal, you might bite one of the few creatures that rabies does not replicate well in, such as a rabbit. So although the behavioral effects of infecting the brain are quite dazzling, if the parasite's impact is too broad, it can wind up in a dead-end host.

Which brings us to a beautifully specific case of brain control and the paper I mentioned earlier, by Manuel Berdoy and his colleagues at the University of Oxford. Berdoy and his associates study a parasite called *Toxoplasma gondii*. In a toxoplasmic utopia, life consists of a two-host sequence involving rodents and cats. The protozoan gets ingested by a rodent, in which it forms cysts throughout the body, particularly in the brain. The rodent gets eaten by a cat, in which the toxoplasma organism reproduces. The cat sheds the parasite in its feces, which, in one of those circles of life, is nibbled by rodents. The whole scenario hinges on specificity: Cats are the only species in which toxoplasma can sexually reproduce and be shed. Thus, toxoplasma wouldn't want its carrier rodent to get picked off by a hawk or its cat feces ingested by a dung beetle. Mind you, the parasite can infect all sorts of other species; it simply has to wind up in a cat if it wants to spread to a new host.

This potential to infect other species is the reason all those "what to do during pregnancy" books recommend banning the cat and its litter box from the house and warn pregnant women against gardening if there are cats wandering about. If toxoplasma from cat feces gets into a pregnant woman, it can get into the fetus, potentially causing neurological damage. Well-informed pregnant women get skittish around cats. Toxoplasma-infected rodents, however, have the opposite reaction. The parasite's extraordinary trick has been to make rodents lose their skittishness.

All good rodents avoid cats — a behavior ethologists call a fixed-action pattern, in that the rodent doesn't develop the aversion because of trial and error (since there aren't likely to be many opportunities to learn from one's errors around cats). Instead feline phobia is hard-wired. And it is accomplished through olfaction in the form of pheromones, the chemical odorant signals that animals release. Rodents instinctually shy away from the smell of a cat — even rodents that have never seen a cat in their lives, rodents that are the descendants of hundreds of generations of lab animals. Except for those infected with toxoplasma. As Berdoy and his group have shown, those rodents selectively lose their aversion to, and fear of, cat pheromones.

Now, this is not some generic case of a parasite messing with the head of the intermediate host and making it scatterbrained and vulnerable. Everything else seems pretty intact in the rodents. The social status of the animal doesn't change in its dominance hierarchy. It is still interested in mating and thus, de facto, in the pheromones of the opposite sex. The infected rodents can still distinguish other odors. They simply don't recoil from cat pheromones. This is flabbergasting. This is akin to someone getting infected with a brain parasite that has no effect whatsoever on the person's thoughts, emotions, SAT scores, or television preferences but, to complete its life cycle, generates an irresistible urge to go to the zoo, scale a fence, and try to French-kiss the pissiest-looking polar bear. A parasite-induced fatal attraction, as Berdoy's team noted in the title of its paper.

Obviously, more research is needed. I say this not only because it is obligatory at this point in any article about science, but because this finding is just so intrinsically cool that someone has to figure out how it works. And because — permit me a Stephen Jay Gould moment — it provides ever more evidence that evolution is amazing. Amazing in ways that are counterintuitive. Many of us hold the deeply entrenched idea that evolution is directional and progressive: Invertebrates are more primitive than vertebrates, mammals are the most evolved of vertebrates, primates are the genetically fanciest mammals, and so forth. Some of my best students consistently fall for that one, no matter how much I drone on in lectures. If you buy into that idea big-time, you're not just wrong, you're not all that many steps away from a philosophy that has humans directionally evolved as well, with the most evolved being northern Europeans with a taste for schnitzel and goose-stepping.

So remember, creatures are out there that can control brains. Microscopic and even larger organisms that have more power than Big Brother and, yes, even neuroscientists. My reflection on a curbside puddle brought me to the opposite conclusion that Narcissus reached in his watery reflection. We need phylogenetic humility. We are certainly not the most evolved species around, nor the least vulnerable. Nor the cleverest.

ERIC SCIGLIANO

# Through the Eye of an Octopus

FROM *Discover*

WHEN BIOLOGIST Roland Anderson of the Seattle Aquarium pulled back the tank's lid, I wasn't sure whether it was to let me get a look at Steve or to let Steve get a look at me. Clearly, Steve was looking — his big hooded eye followed me, and a single five-foot-long arm reached out to the hand I held above the water's surface. The arm inched up past my wrist to my shoulder, its suckers momentarily attaching and releasing like cold kisses. I couldn't help feeling as if I was being tasted, and I was, by tens of thousands of chemoreceptors. And I couldn't help feeling as if I were being studied, that a measuring intelligence lay behind that intent eye and exploring arm.

Finally, when the arm's fingerlike tip reached my neck, it shot back like a snapped rubber band. Steve curled into a tight, defensive ball in the corner of the tank. His skin texture changed from glassy smooth to a fissured moonscape; his color changed from mottled brown to livid red — which seemed to signal anger — and he squinted at me. Had something alarmed or offended him? Perhaps we were both a great mystery to each other.

Octopuses and their cephalopod cousins the cuttlefish and the squid are evolutionary oxymorons: big-brained invertebrates that display many cognitive, behavioral, and affective traits once considered exclusive to the higher vertebrates. They challenge the deep-seated notion that intelligence advanced from fish and amphibians to reptiles, birds, mammals, early primates, and finally humans. These are mollusks, after all — cousins to brainless clams and oysters, passive filter feeders that get along just fine, thank you, with a

few ganglia for central nervous systems. Genetic studies show that
mollusk ancestors split from the vertebrates around 1.2 billion
years ago, making humans at least as closely related to shrimps,
starfish, and earthworms as to octopuses. And so questions loom:
How could asocial invertebrates with short life spans develop signs
of intelligence? And why?

Although biologists are just beginning to probe these questions,
those who observe the creatures in their natural haunts have long
extolled their intelligence. "Mischief and craft are plainly seen to
be the characteristics of this creature," the Roman natural historian Claudius Aelianus wrote at the turn of the third century A.D.
Today's divers marvel at the elaborate trails the eight-leggers follow
along the seafloor, and at their irrepressible curiosity: Instead of
fleeing, some octopuses examine divers the way Steve checked me
out, tugging at their masks and air regulators. Researchers and
aquarium attendants tell tales of octopuses that have tormented
and outwitted them. Some captive octopuses lie in ambush and spit
in their keepers' faces. Others dismantle pumps and block drains,
causing costly floods, or flex their arms in order to pop locked lids.
Some have been caught sneaking from their tanks at night into
other exhibits, gobbling up fish, then sneaking back to their tanks,
damp trails along walls and floors giving them away.

That Steve was named Steve was also revealing: Octopuses are
the only animals, other than mammals like cuddly seals, that aquarium workers bother to name. So Anderson, Seattle's lead invertebrate biologist, began to wonder: If keepers recognize octopuses as
individuals, how much difference is there among individual octopuses? Might these bizarre-looking mollusks have personalities?
And if so, how else might their evolution have converged with ours
across a billion-year chasm?

Meanwhile, in the waters off Bermuda, Canadian comparative
psychologist Jennifer Mather was asking similar questions. Mather
had observed an *Octopus vulgaris,* the common Atlantic octopus,
catch several crabs and return to its rock den to eat them. Afterward it emerged, gathered four stones, propped these at the den
entrance, and, thus shielded, took a safe siesta. The strategy suggested qualities that weren't supposed to occur in the lower orders:
foresight, planning, perhaps even tool use.

When Mather and Anderson met at a conference, they discovered they had stumbled onto similar phenomena and began collaborating. Other scientists had already tested the ability of octopuses to solve mazes, learn cues, and remember solutions. They had found that octopuses solve readily, learn quickly, and, in the short term, remember what they have learned. Mather and Anderson delved deeper, documenting a range of qualities and activities closely associated with intelligence but previously known only in advanced vertebrates. Some of their work has been controversial, and some of their conclusions have been disputed. But other researchers are now confirming their key points and logging even more startling findings.

Anatomy confirms what behavior reveals: Octopuses and cuttlefish have larger brains, relative to body weight, than most fish and reptiles, larger on average than any animals save birds and mammals. Although an octopus brain differs from a typical vertebrate's brain — it wraps around the esophagus instead of resting in a cranium — it also shares key features such as folded lobes, a hallmark of complexity, and distinct visual and tactile memory centers. It even generates similar electrical patterns. Electroencephalograms of other invertebrates show spiky static — "like bacon frying," says neurophysiologist Ted Bullock of the University of California at San Diego, who nonetheless found vertebratelike slow waves in octopuses and cuttlefish. The pattern, he says, is "similar to but weaker than that of a dog, a dolphin, a human."

Researchers at the Konrad Lorenz Institute for Evolution and Cognition Research in Austria recently found one more telling indicator: Octopuses, which rely on monocular vision, favor one eye over the other. Such lateralization, corresponding to our right- and left-handedness, suggests specialization in the brain's hemispheres, which is believed to improve its efficiency and which was first considered an exclusively human, then an exclusively vertebrate, attribute.

The mystery deepens. According to the social theory of intelligence articulated by N. K. Humphrey and Jane Goodall, complex brains blossom in complex social settings; chimps and dolphins have to be smart to read the intentions of other chimps and dolphins. Moreover, such smarts arise in long-lived animals: Extended childhoods and parental instruction enable them to learn, and longevity justifies the investment in big brains. But many cephalopods live less than a year, and the giant Pacific octopus, which has one of the longest documented life spans, survives for only four years. Their social lives are simple to nonexistent: Squid form schools, but they don't seem to establish individual relationships. Cuttlefish gather while young and later on to mate, but they don't form social structures. Octopuses are solitary; they breed once, then waste away and die. Females tend their eggs, but the tiny hatchlings are on their own. As cephalopod-respiration expert Ron O'Dor of Dalhousie University in Nova Scotia wonders, "Why would you bother to get so smart when you're so short-lived?"

For Jennifer Mather, pursuing those questions marks a convergence of childhood and adult passions. Mather grew up in Victoria, British Columbia, along a biologically rich shoreline. "I got fascinated with intertidal life," she recalls. "I always thought I'd study mollusks." In college, she took an animal-behavior class and had an epiphany: "Most people in comparative psychology compare humans and other primates," she observes, which leaves the field wide open for studies of mollusk behavior and cognition. "And if you talk about mollusk behavior, you're talking about cephalopods."

Mather landed in an unlikely spot for marine research: at the University of Lethbridge in landlocked Alberta, which hasn't had any cephalopods since the Devonian Period. But in the 1980s academic jobs were scarce. Mather then found a lab with Anderson in Seattle and a field base near a secluded coral reef off Bonaire, an island in the Netherlands Antilles. There she leads an international investigation of communication and interactions among Caribbean reef squid — the first long-term study of a wild cephalopod population.

In Seattle, Mather and Anderson have pursued octopuses. Perhaps their most startling and controversial finding is that individuals show distinct personality traits, the first ever measured in an invertebrate. They found that octopuses confronted with the same threat alerts and food stimuli react in different ways. One might flee, but another might fight or show curiosity. That sets them apart from other invertebrates, says Shelley Adamo, a psychologist at Dalhousie who has studied both cephalopods and insects. For example, individual crickets may behave differently at different times — singing today and silent tomorrow. But they don't have consistent patterns that set one cricket apart from another.

*Personality* can be a controversial word. Some behavioralists call such labels anthropomorphic, while others contend that it's anthropocentric to presume other animals cannot have personalities. Some of Mather and Anderson's peers feel more comfortable with the findings than the terminology. "They do good work and ask interesting questions," says cephalopod researcher John Cigliano of Cedar Crest College in Allentown, Pennsylvania. "But I'm not entirely convinced. It's a tricky business just coming up with a definition of personality." David Sinn, a graduate student at Portland State University, followed up Mather and Anderson's personality work with a more extensive study that they coauthored. That study avoided the "p" word, charting the emergence of key "temperamental traits" in seventy-three lab-bred California octopuses. It found considerable temperamental variation and distinct developmental stages. Like mammals, Sinn's octopuses were more active and aggressive when young and grew more alert to danger as they matured — evidence that their behavior was learned.

Previous researchers tested octopuses in artificial mazes; Mather and Anderson found ways to observe learning and cognition in more natural circumstances. They charted the efficiency and flexibility with which giant Pacific octopuses switch strategies to open different shellfish — smashing thin mussels, prying open clams, drilling tougher-shelled clams with their rasplike radulae. When served clams sealed with steel wire, for example, octopuses deftly switched from prying to drilling.

Tool use was once commonly invoked as uniquely human. Scientists know better now, but they still cite it as evidence of distinguishing intelligence in chimpanzees, elephants, and crows. Mather describes several ways octopuses use their water jets as tools: to clean their dens, push away rocks and other debris, and drive off pesky scavenger fish.

In 1999 she and Anderson published an even more sensational claim: that octopuses engage in play, the deliberate, repeated, outwardly useless activity through which smarter animals explore their world and refine their skills. Amateur aquarists were the first to suspect that octopuses played. While still in high school, James Wood, now a marine biologist at the University of Texas's marine lab in Galveston, watched his pet octopus grab, submerge, and release her tank's floating hydrometer as if she were a toddler with a bath toy. She also spread her mantle and "bubble surfed" the tank's aerator jets.

Anderson tested for play by presenting eight giant Pacific octopuses with floating pill bottles in varying colors and textures twice a day for five days. Six octopuses examined the bottles and lost interest, but two blew them repeatedly into their tanks' jets. One propelled a bottle at an angle so it circled the tank; the other shot it so it rebounded quickly — and on three occasions shot it back at least twenty times, as if it were bouncing a ball.

One respected cephalopod expert isn't convinced. Jean Boal, an animal behaviorist at Millersville University in Pennsylvania, is acutely aware of the dangers of getting carried away when studying these charismatic megamollusks. She previously worked at the Zoological Station in Naples, a wellspring of cephalopod research. In 1992 Graziano Fiorito, a researcher at that lab, announced a bombshell: Octopuses could learn by watching other octopuses. Such observational learning, a hallmark of intelligent social animals, seemed impossible. And it probably was. Other researchers, including Boal, have been unable to reproduce Fiorito's results. Some questioned his methodology, and for a year or two the controversy cast a pall on research into octopus learning.

Boal subsequently withdrew her own initial findings of complex learning by octopuses. She has since carved herself a niche as the field's designated skeptic, often questioning conclusions and urging more rigor. "My bias is to build a case slowly, with careful science," Boal says quietly. "That's not the case with all cephalopod biologists." She doesn't rule out the possibility that octopuses play, but she questions whether the bottle-jetters did: "It could reflect boredom, like a cat pacing."

One authority on play behavior, psychologist Gordon Burghardt at the University of Tennessee in Knoxville, says that as Anderson and Mather describe it, the bottle-jetting would qualify as play. Boredom, he says, can be "a trigger for play." And other confirmation is emerging. Doubting the Seattle findings, Ulrike Griebel of the Lorenz Institute recently conducted more extensive trials. She offered common octopuses varied objects, from Lego assemblies to floating bottles on strings (a favorite). Some octopuses took toys into their nests and toted them along while fetching food — acquisitive behavior that Griebel says "might be an early stage of object play."

Meanwhile, Anderson has been investigating another phenomenon little noted in invertebrates: sleep. Until recently, only vertebrates were believed to sleep in the full metabolic sense. But Anderson has observed that octopuses, ordinarily hypervigilant, may sleep deeply. Their eyes glaze over, their breathing turns slow and shallow, they don't respond to light taps, and a male will let his delicate ligula — the sex organ at the tip of one arm — dangle perilously.

Stephen Duntley, a sleep specialist at Washington University Medical School in St. Louis, has videotaped similar slumber in cuttlefish, with a twist: Sleeping cuttlefish lie still, their skin a dull brown, for ten- to fifteen-minute stretches, then flash bold colored patterns and twitch their tentacles for briefer intervals. After viewing Duntley's footage, Anderson suggests the cuttlefish might merely be waking to check for threats. But Duntley says the cycling resembles the rapid-eye-movement sleep of birds and mammals, when humans dream. If invertebrates undergo a similar cycle, Duntley argues, it would affirm "that REM sleep is very important to learning." Would it also suggest that cuttlefish and octopuses dream? "That's the ultimate question," Duntley responds.

The ultimate question, with octopuses as with other sentient creatures, may be how we should treat them. In 2001 Mather argued in *The Journal of Applied Welfare Science* that people should err on the humane side, since some octopuses "very likely have the capacity for pain and suffering and, perhaps, mental suffering." If captive cephalopods suffer mentally — or even get "bored," as Boal puts it — then they should benefit from enrichment: amenities and activities that replicate elements of their natural environment. Mather, Anderson, and Wood have urged enriched environments but have no experimental evidence that it makes a difference. Recently that evidence came from a French study that even the skeptical Boal calls "beautiful work." Ludovic Dickel, a neuroethologist at the University of Caen, found that cuttlefish raised in groups and in tanks with sand, rocks, and plastic seaweed grew faster, learned faster, and retained more of what they learned than those raised alone in bare tanks. Performance rose in animals transferred midway from impoverished to enriched conditions and declined in those transferred to solitary confinement.

Other evidence suggests that solitary octopuses, like solitary orangutans, may communicate more with others of their species than researchers previously realized. Cigliano found that California octopuses that were kept together quickly established hierarchies and avoided wasteful, dangerous confrontations; the weaker animals seemed to recognize and yield to the stronger ones, even when the latter were hidden in their dens. The flip side of communication is deception, another hallmark of intelligence. And some octopuses and cuttlefish practice it. Male cuttlefish adopt female coloring, patterns, and shape — to mate surreptitiously with females guarded by larger rivals. And the Indonesian mimic octopus fools predators by impersonating poisonous soles and venomous lionfish, sea snakes, and possibly jellyfish and sea anemones.

And so, piece by piece, Mather, Anderson, and other researchers fill in the puzzle. A picture emerges of convergent evolution across a billion-year gap. One after another, these precocious invertebrates display what were supposed to be special traits of advanced vertebrates. But one question nags: Why would short-lived, solitary creatures acquire so many of the cognitive and affective features of long-lived, social vertebrates?

Mather proposes "a foraging theory of intelligence." She says that animals like octopuses (or humans) that pursue varied food sources in changeable, perilous habitats must develop a wide range of hunting and defensive strategies. That takes brainpower. "If you find yourself foraging in a complex environment, where you have to deal with many kinds of prey and predators," she says, "it makes sense to invest more in cognition." Temperamental variation — call it personality — also helps a species survive in a volatile, supercompetitive milieu by ensuring that different individuals respond differently to changing conditions, so some will thrive. Even semelparity, the live-fast-die-young strategy of growing quickly and throwing everything into one breeding blast, may serve that end by assuring rapid turnover and regeneration.

Although cephalopods are an ancient order, shell-less cephalopods are relatively recent arrivals — about 200 million years old, like mammals and teleost, or bony, fishes. Before that, ammonites and other shelled cephalopods ruled the seas, but competition from the nimble, fast-swimming teleosts wiped out all but the relic nautilus. The cephalopods that survived were the zoological counterrevolutionaries that turned the vertebrates' weapons against them. They shed their shells and became speedy, like squid, or they became clever and elusive, like octopuses and cuttlefish. Octopuses, naked and vulnerable, took to dens, as early humans took to caves. Like humans, they became versatile foragers, using a wide repertoire of stalking and killing techniques. To avoid exposure, they developed spatial sense and learned to cover their hunting grounds methodically and efficiently. Mather and O'Dor found that the Bermudan *O. vulgaris* spends just 7 percent of its time hunting; Australian giant cuttlefish spend 3 percent.

In short, octopuses came to resemble us. Their hunting done, they huddle safely in their dens, a bit like early humans around campfires. "You have to wonder what they think about while they're tucked away," says O'Dor. Do they muse on the cruel turns of evolution, which have left them all dressed up with big brains but with no place to go and little time to use them?

MEREDITH F. SMALL

# *Captivated*

FROM *Natural History*

I'M SITTING ON A BENCH in New York City's Central Park, waiting for the zoo to open. I have spent years observing macaque monkeys in the field, but these days I only teach and write about what they do, and I miss them. So whenever I'm in Manhattan, I hang out here with the snow monkeys *(Macaca fuscata).*

I've been visiting this troop for years. I have seen them in sunshine and snow; stood in the rain and watched them lick drops of wetness off their fur; held short business meetings in front of their exhibit; forced friends to meet me here. Unbeknownst to them, these furry gray monkeys from Japan have become my primate touchstone.

On this visit it's clear and sunny, and through the entrance gates I see the macaques jumping around their island exhibit. A path of rocks breaks the surface of the retaining pond that surrounds their enclosure, and a young female hops from one to another, leapfrogging over her troopmates as she goes.

Finally the gates open, and as I approach the group, my professional observing skills click in. By the time I reach them, my training as an observer — and that touch of magic I always feel in the presence of monkeys — has locked out the world; all that matters is the movement of these animals.

Today I count nine adults, one juvenile, and no babies. I know that fall is breeding season, and the females are signaling their fertility with red behinds. To my right a status interaction is unfolding — a female turns her rear to another female, indicating her lower position. I lean across the rail and get into the Zen of figuring out what these monkeys already know about each other — who is related to whom, how their rank is doled out, who will mate next.

My primatological reverie is interrupted by a crowd of visitors. I hear one woman call a male "she," and I'm compelled to correct her. "It's the shape of his face," I tell her, "and his size — and those bright red testicles." But I should know better than to be so patronizing, such a know-it-all. Several years ago, on one frozen January day, I asked some of the zoo's wild-animal keepers why the snow monkeys were indoors. After all, I told them, these monkeys are accustomed to crawling through snowdrifts in their native Japan. "If the pond froze over," they patiently responded, "the monkeys would simply walk out of the zoo." Humbled, I went to see the polar bear.

When I have the monkeys to myself again, I walk up the hill behind the exhibit and lean over the granite wall overlooking their enclosure, focusing on a pair of females. One is stretched out on a rock, arms and legs splayed in relaxation. Her eyelids droop. She is at peace. The other methodically moves a hand across her partner's belly, separating each strand of hair, gently touching each exposed patch of skin. Monkeys have done this to me, sitting on my shoulders with their handlike feet pressed against my neck, picking through my hair. I know it feels like heaven.

Concentrating on the grooming females, I stretch my own arms across the wall and feel the reflected warmth of the sun seep up from the granite slab. I, too, let my eyelids droop in contentment. For a few precious minutes I pretend that I have done nothing for the past few months but watch this group, that we know each other intimately, observer and observed. Monkey noises, their barks and calls, fill my ears. The familiar, musty odor of monkey fur at close quarters fills my nostrils.

I am, once again, renewed.

MAX TEGMARK

# Parallel Universes

FROM *Scientific American*

IS THERE A COPY of you reading this article? A person who is not you but who lives on a planet called Earth, with misty mountains, fertile fields and sprawling cities, in a solar system with eight other planets? The life of this person has been identical to yours in every respect. But perhaps he or she now decides to put down this article without finishing it, while you read on.

The idea of such an alter ego seems strange and implausible, but it looks as if we will just have to live with it, because it is supported by astronomical observations. The simplest and most popular cosmological model today predicts that you have a twin in a galaxy about 10 to the $10^{28}$ meters from here. This distance is so large that it is beyond astronomical, but that does not make your doppelgänger any less real. The estimate is derived from elementary probability and does not even assume speculative modern physics, merely that space is infinite (or at least sufficiently large) in size and almost uniformly filled with matter, as observations indicate. In infinite space, even the most unlikely events must take place somewhere. There are infinitely many other inhabited planets, including not just one but infinitely many that have people with the same appearance, name, and memories as you, who play out every possible permutation of your life choices.

You will probably never see your other selves. The farthest you can observe is the distance that light has been able to travel during the 14 billion years since the big bang expansion began. The most distant visible objects are now about $4 \times 10^{26}$ meters away — a distance that defines our observable universe, also called our Hubble volume, our horizon volume, or simply our universe. Likewise, the universes of your other selves are spheres of the same size centered on their planets. They are the most straightforward example of parallel universes. Each universe is merely a small part of a larger "multiverse."

By this very definition of "universe," one might expect the notion of a multiverse to be forever in the domain of metaphysics. Yet the borderline between physics and metaphysics is defined by whether a theory is experimentally testable, not by whether it is weird or involves unobservable entities. The frontiers of physics have gradually expanded to incorporate ever more abstract (and once metaphysical) concepts such as a round Earth, invisible electromagnetic fields, time slowdown at high speeds, quantum superpositions, curved space, and black holes. Over the past several years the concept of a multiverse has joined this list. It is grounded in well-tested theories such as relativity and quantum mechanics, and it fulfills both of the basic criteria of an empirical science: it makes predictions, and it can be falsified. Scientists have discussed as many as four distinct types of parallel universes. They key question is not whether the multiverse exists but rather how many levels it has.

## Level I: Beyond Our Cosmic Horizon

The parallel universes of your alter egos constitute the Level I multiverse. It is the least controversial type. We all accept the existence of things that we cannot see but could see if we moved to a different vantage point or merely waited, like people watching for ships to come over the horizon. Objects beyond the cosmic horizon have a similar status. The observable universe grows by a light-year every year as light from farther away has time to reach us. An infinity lies out there, waiting to be seen. You will probably die long before your alter egos come into view, but in principle, and if cosmic expansion cooperates, your descendants could observe them through a sufficiently powerful telescope.

If anything, the Level I multiverse sounds trivially obvious. How could space *not* be infinite? Is there a sign somewhere saying "Space Ends Here — Mind the Gap"? If so, what lies beyond it? In fact, Einstein's theory of gravity calls this intuition into question. Space could be finite if it has a convex curvature or an unusual topology (that is, interconnectedness). A spherical, doughnut-shaped, or pretzel-shaped universe would have a limited volume and no edges. The cosmic microwave background radiation allows sensitive tests of such scenarios. So far, however, the evidence is against them. Infinite models fit the data, and strong limits have been placed on the alternatives.

Another possibility is that space is infinite but matter is confined to a finite region around us — the historically popular "island universe" model. In a variant on this model, matter thins out on large scales in a fractal pattern. In both cases, almost all universes in the Level I multiverse would be empty and dead. But recent observations of the three-dimensional galaxy distribution and the microwave background have shown that the arrangement of matter gives way to dull uniformity on large scales, with no coherent structures larger than about $10^{24}$ meters. Assuming that this pattern continues, space beyond our observable universe teems with galaxies, stars, and planets.

Observers living in Level I parallel universes experience the same laws of physics as we do but with different initial conditions. According to current theories, processes early in the big bang spread matter around with a degree of randomness, generating all possible arrangements with nonzero probability. Cosmologists assume that our universe, with an almost uniform distribution of matter and initial density fluctuations of one part in 100,000, is a fairly typical one (at least among those that contain observers). That assumption underlies the estimate that your closest identical copy is 10 to the $10^{28}$ meters away. About 10 to the $10^{92}$ meters away, there should be a sphere of radius 100 light-years identical to the one centered here, so all perceptions that we have during the next century will be identical to those of our counterparts over there. About 10 to the $10^{118}$ meters away should be an entire Hubble volume identical to ours.

These are extremely conservative estimates, derived simply by counting all possible quantum states that a Hubble volume can have if it is no hotter than $10^8$ kelvins. One way to do the calculation is to ask how many protons could be packed into a Hubble volume at that temperature. The answer is $10^{118}$ protons. Each of those particles may or may not, in fact, be present, which makes for 2 to the $10^{118}$ possible arrangements of protons. A box containing that many Hubble volumes exhausts all the possibilities. If you round off the numbers, such a box is about 10 to the $10^{118}$ meters across. Beyond that box, universes — including ours — must repeat. Roughly the same number could be derived by using thermodynamic or quantum-gravitational estimates of the total information content of the universe.

Your nearest doppelgänger is most likely to be much closer than these numbers suggest, given the processes of planet formation and biological evolution that tip the odds in your favor. Astronomers suspect that our Hubble volume has at least $10^{20}$ habitable planets; some might well look like Earth.

The Level I multiverse framework is used routinely to evaluate theories in modern cosmology, although this procedure is rarely spelled out explicitly. For instance, consider how cosmologists used the microwave background to rule out a finite spherical geometry. Hot and cold spots in microwave background maps have a characteristic size that depends on the curvature of space, and the observed spots appear too small to be consistent with a spherical shape. But it is important to be statistically rigorous. The average spot size varies randomly from one Hubble volume to another, so it is possible that our universe is fooling us — it could be spherical but happen to have abnormally small spots. When cosmologists say they have ruled out the spherical model with 99.9 percent confidence, they really mean that if this model were true, fewer than one in 1,000 Hubble volumes would show spots as small as those we observe.

The lesson is that the multiverse theory can be tested and falsified even though we cannot see the other universes. The key is to predict what the ensemble of parallel universes is and to specify a probability distribution, or what mathematicians call a "measure," over that ensemble. Our universe should emerge as one of the most probable. If not — if, according to the multiverse theory, we live in an improbable universe — then the theory is in trouble. As I will discuss later, this measure problem can become quite challenging.

## Level II: Other Postinflation Bubbles

If the level I multiverse was hard to stomach, try imagining an infinite set of distinct Level I multiverses, some perhaps with different space-time dimensionality and different physical constants. Those other multiverses — which constitute a Level II multiverse — are predicted by the currently popular theory of chaotic eternal inflation.

Inflation is an extension of the big bang theory and ties up many of the loose ends of that theory, such as why the universe is so big, so uniform, and so flat. A rapid stretching of space long ago can explain all these and other attributes in one fell swoop. Such stretching is predicted by a wide class of theories of elementary particles, and all available evidence bears it out. The phrase "chaotic eternal" refers to what happens on the very largest scales. Space as a whole is stretching and will continue doing so forever, but some regions of space stop stretching and form distinct bubbles, like gas pockets in a loaf of rising bread. Infinitely many such bubbles emerge. Each is an embryonic Level I multiverse: infinite in size and filled with matter deposited by the energy field that drove inflation.

Those bubbles are more than infinitely far away from Earth, in the sense that you would never get there even if you traveled at the speed of light forever. The reason is that the space between our bubble and its neighbors is expanding faster than you could travel through it. Your descendants will never see their doppelgängers elsewhere in Level II. For the same reason, if cosmic expansion is accelerating, as observations now suggest, they might not see their alter egos even in Level I.

The Level II multiverse is far more diverse than the Level I multiverse. The bubbles vary not only in their initial conditions but also in seemingly immutable aspects of nature. The prevailing view in physics today is that the dimensionality of space-time, the qualities of elementary particles, and many of the so-called physical constants are not built into physical laws but are the outcome of processes known as symmetry breaking. For instance, theorists think that the space in our universe once had nine dimensions, all on an equal footing. Early in cosmic history, three of them partook in the cosmic expansion and became the three dimensions we now observe. The other six are now unobservable, either because they have stayed microscopic with a doughnutlike topology or because all matter is confined to a three-dimensional surface (a membrane, or simply "brane") in the nine-dimensional space.

Thus, the original symmetry among the dimensions broke. The

quantum fluctuations that drive chaotic inflation could cause different symmetry breaking in different bubbles. Some might become four-dimensional, others could contain only two rather than three generations of quarks, and still others might have a stronger cosmological constant than our universe does.

Another way to produce a Level II multiverse might be through a cycle of birth and destruction of universes. In a scientific context, this idea was introduced by physicist Richard C. Tolman in the 1930s and recently elaborated on by Paul J. Steinhardt of Princeton University and Neil Turok of the University of Cambridge. The Steinhardt and Turok proposal and related models involve a second three-dimensional brane that is quite literally parallel to ours, merely offset in a higher dimension. This parallel universe is not really a separate universe, because it interacts with ours. But the ensemble of universes — past, present, and future — that these branes create would form a multiverse, arguably with a diversity similar to that produced by chaotic inflation. An idea proposed by physicist Lee Smolin of the Perimeter Institute in Waterloo, Ontario, involves yet another multiverse comparable in diversity to that of Level II but mutating and sprouting new universes through black holes rather than through brane physics.

Although we cannot interact with other Level II parallel universes, cosmologists can infer their presence indirectly, because their existence can account for unexplained coincidences in our universe. To give an analogy, suppose you check into a hotel, are assigned room 1967, and note that this is the year you were born. What a coincidence, you say. After a moment of reflection, however, you conclude that this is not so surprising after all. The hotel has hundreds of rooms, and you would not have been having these thoughts in the first place if you had been assigned one with a number that meant nothing to you. The lesson is that even if you knew nothing about hotels, you could infer the existence of other hotel rooms to explain the coincidence.

As a more pertinent example, consider the mass of the sun. The mass of a star determines its luminosity, and using basic physics, one can compute that life as we know it on Earth is possible only if the sun's mass falls into the narrow range between $1.6 \times 10^{30}$ and $2.4 \times 10^{30}$ kilograms. Otherwise Earth's climate would be colder than that of present-day Mars or hotter than that of present-day

Venus. The measured solar mass is $2.0 \times 10^{30}$ kilograms. At first glance, this apparent coincidence of the habitable and observed mass values appears to be a wild stroke of luck. Stellar masses run from $10^{29}$ to $10^{32}$ kilograms, so if the sun acquired its mass at random, it had only a small chance of falling into the habitable range. But just as in the hotel example, one can explain this apparent coincidence by postulating an ensemble (in this case, a number of planetary systems) and a selection effect (the fact that we must find ourselves living on a habitable planet). Such observer-related selection effects are referred to as "anthropic," and although the "A-word" is notorious for triggering controversy, physicists broadly agree that these selection effects cannot be neglected when testing fundamental theories.

What applies to hotel rooms and planetary systems applies to parallel universes. Most, if not all, of the attributes set by symmetry breaking appear to be fine-tuned. Changing their values by modest amounts would have resulted in a qualitatively different universe — one in which we probably would not exist. If protons were 0.2 percent heavier, they could decay into neutrons, destabilizing atoms. If the electromagnetic force were 4 percent weaker, there would be no hydrogen and no normal stars. If the weak interaction were much weaker, hydrogen would not exist; if it were much stronger, supernovae would fail to seed interstellar space with heavy elements. If the cosmological constant were much larger, the universe would have blown itself apart before galaxies could form.

Although the degree of fine-tuning is still debated, these examples suggest the existence of parallel universes with other values of the physical constants. The Level II multiverse theory predicts that physicists will never be able to determine the values of these constants from first principles. They will merely compute probability distributions for what they should expect to find, taking selection effects into account. The result should be as generic as is consistent with our existence.

## Level III: Quantum Many Worlds

The Level I and Level II multiverses involve parallel worlds that are far away, beyond the domain even of astronomers. But the next level of multiverse is right around you. It arises from the famous, and famously controversial, many-worlds interpretation of quantum mechanics — the idea that random quantum processes cause the universe to branch into multiple copies, one for each possible outcome.

In the early twentieth century the theory of quantum mechanics revolutionized physics by explaining the atomic realm, which does not abide by the classical rules of Newtonian mechanics. Despite the obvious successes of the theory, a heated debate rages about what it really means. The theory specifies the state of the universe not in classical terms, such as the positions and velocities of all particles, but in terms of a mathematical object called a wave function. According to the Schrödinger equation, this state evolves over time in a fashion that mathematicians term "unitary," meaning that the wave function rotates in an abstract infinite-dimensional space called Hilbert space. Although quantum mechanics is often described as inherently random and uncertain, the wave function evolves in a deterministic way. There is nothing random or uncertain about it.

The sticky part is how to connect this wave function with what we observe. Many legitimate wave functions correspond to counterintuitive situations, such as a cat being dead and alive at the same time in a so-called superposition. In the 1920s physicists explained away this weirdness by postulating that the wave function "collapsed" into some definite classical outcome whenever someone made an observation. This add-on had the virtue of explaining observations, but it turned an elegant, unitary theory into a kludgy, nonunitary one. The intrinsic randomness commonly ascribed to quantum mechanics is the result of this postulate.

Over the years many physicists have abandoned this view in favor of one developed in 1957 by Princeton graduate student Hugh Everett III. He showed that the collapse postulate is unnecessary. Unadulterated quantum theory does not, in fact, pose any contradictions. Although it predicts that one classical reality gradually splits into superpositions of many such realities, observers subjectively experience this splitting merely as a slight randomness, with probabilities in exact agreement with those from the old collapse postulate. This superposition of classical worlds is the Level III multiverse.

Everett's many-worlds interpretation has been boggling minds inside and outside physics for more than four decades. But the theory becomes easier to grasp when one distinguishes between two ways of viewing a physical theory: the outside view of a physicist studying its mathematical equations, like a bird surveying a landscape from high above it, and the inside view of an observer living in the world described by the equations, like a frog living in the landscape surveyed by the bird.

From the bird perspective, the Level III multiverse is simple. There is only one wave function. It evolves smoothly and deterministically over time without any kind of splitting or parallelism. The abstract quantum world described by this evolving wave function contains within it a vast number of parallel classical story lines, continuously splitting and merging, as well as a number of quantum phenomena that lack a classical description. From their frog perspective, observers perceive only a tiny fraction of this full reality. They can view their own Level I universe, but a process called decoherence — which mimics wave function collapse while preserving unitarity — prevents them from seeing Level III parallel copies of themselves.

Whenever observers are asked a question, make a snap decision, and give an answer, quantum effects in their brains lead to a superposition of outcomes, such as "Continue reading the article" and "Put down the article." From the bird perspective, the act of making a decision causes a person to split into multiple copies: one who keeps on reading and one who doesn't. From their frog perspective, however, each of these alter egos is unaware of the others and notices the branching merely as a slight randomness: a certain probability of continuing to read or not.

As strange as this may sound, the exact same situation occurs even in the Level I multiverse. You have evidently decided to keep on reading the article, but one of your alter egos in a distant galaxy put down the magazine after the first paragraph. The only difference between Level I and Level III is where your doppelgängers reside. In Level I they live elsewhere in good old three-dimensional space. In Level III they live on another quantum branch in infinite-dimensional Hilbert space.

The existence of Level III depends on one crucial assumption: that the time evolution of the wave function is unitary. So far experimenters have encountered no departures from unitarity. In the past few decades they have confirmed unitarity for ever larger systems, including carbon 60 buckyball molecules and kilometer-long optical fibers. On the theoretical side, the case for unitarity has been bolstered by the discovery of decoherence. Some theorists who work on quantum gravity have questioned unitarity; one concern is that evaporating black holes might destroy information, which would be a nonunitary process. But a recent breakthrough in string theory known as AdS/CFT correspondence suggests that even quantum gravity is unitary. If so, black holes do not destroy information but merely transmit it elsewhere.

If physics is unitary, then the standard picture of how quantum fluctuations operated early in the big bang must change. These fluctuations did not generate initial conditions at random. Rather they generated a quantum superposition of all possible initial conditions, which coexisted simultaneously. Decoherence then caused these initial conditions to behave classically in separate quantum branches. Here is the crucial point: The distribution of outcomes on different quantum branches in a given Hubble volume (Level III) is identical to the distribution of outcomes in different Hubble volumes within a single quantum branch (Level I). This property of the quantum fluctuations is known in statistical mechanics as ergodicity.

The same reasoning applies to Level II. The process of symmetry breaking did not produce a unique outcome but rather a superposition of all outcomes, which rapidly went their separate ways. So if physical constants, space-time dimensionality, and so on can vary among parallel quantum branches at Level III, then they will also vary among parallel universes at Level II.

In other words, the Level III multiverse adds nothing new beyond Level I and Level II, just more indistinguishable copies of the same universes — the same old story lines playing out again and again in other quantum branches. The passionate debate about Everett's theory therefore seems to be ending in a grand anticlimax, with the discovery of less controversial multiverses (Levels I and II) that are equally large.

Needless to say, the implications are profound, and physicists are only beginning to explore them. For instance, consider the ramifications of the answer to a long-standing question: Does the number of universes exponentially increase over time? The surprising answer is no. From the bird perspective, there is of course only one quantum universe. From the frog perspective, what matters is the

number of universes that are distinguishable at a given instant —
that is, the number of noticeably different Hubble volumes. Imagine moving planets to random new locations, imagine having married someone else, and so on. At the quantum level, there are 10 to
the $10^{118}$ universes with temperatures below $10^8$ kelvins. That is a
vast number, but a finite one.

From the frog perspective, the evolution of the wave function
corresponds to a never-ending sliding from one of these 10 to the
$10^{118}$ states to another. Now you are in universe A, the one in which
you are reading this sentence. Now you are in universe B, the one
in which you are reading this other sentence. Put differently, universe B has an observer identical to one in universe A, except with
an extra instant of memories. All possible states exist at every instant, so the passage of time may be in the eye of the beholder —
an idea explored in Greg Egan's 1994 science-fiction novel *Permutation City* and developed by physicist David Deutsch of the University of Oxford, independent physicist Julian Barbour, and others.
The multiverse framework may thus prove essential to understanding the nature of time.

## Level IV: Other Mathematical Structures

The initial conditions and physical constants in the Level I, Level
II, and Level III multiverses can vary, but the fundamental laws that
govern nature remain the same. Why stop there? Why not allow the
laws themselves to vary? How about a universe that obeys the laws
of classical physics, with no quantum effects? How about time that
comes in discrete steps, as for computers, instead of being continu-
ous? How about a universe that is simply an empty dodecahedron?
In the Level IV multiverse, all these alternative realities actually
exist.

A hint that such a multiverse might not be just some beer-fueled
speculation is the tight correspondence between the worlds of abstract reasoning and of observed reality. Equations and, more generally, mathematical structures such as numbers, vectors, and geo-
metric objects describe the world with remarkable verisimilitude.
In a famous 1959 lecture, physicist Eugene P. Wigner argued that
"the enormous usefulness of mathematics in the natural sciences is
something bordering on the mysterious." Conversely, mathemati-

cal structures have an eerily real feel to them. They satisfy a central criterion of objective existence: They are the same no matter who studies them. A theorem is true regardless of whether it is proved by a human, a computer, or an intelligent dolphin. Contemplative alien civilizations would find the same mathematical structures as we have. Accordingly, mathematicians commonly say that they discover mathematical structures rather than create them.

There are two tenable but diametrically opposed paradigms for understanding the correspondence between mathematics and physics, a dichotomy that arguably goes as far back as Plato and Aristotle. According to the Aristotelian paradigm, physical reality is fundamental and mathematical language is merely a useful approximation. According to the Platonic paradigm, the mathematical structure is the true reality and observers perceive it imperfectly. In other words, the two paradigms disagree on which is more basic, the frog perspective of the observer or the bird perspective of the physical laws. The Aristotelian paradigm prefers the frog perspective, whereas the Platonic paradigm prefers the bird perspective.

As children, long before we had even heard of mathematics, we were all indoctrinated with the Aristotelian paradigm. The Platonic view is an acquired taste. Modern theoretical physicists tend to be Platonists, suspecting that mathematics describes the universe so well because the universe is inherently mathematical. Then all of physics is ultimately a mathematics problem: A mathematician with unlimited intelligence and resources could in principle compute the frog perspective — that is, compute what self-aware observers the universe contains, what they perceive, and what languages they invent to describe their perceptions to one another.

A mathematical structure is an abstract, immutable entity existing outside of space and time. If history were a movie, the structure would correspond not to a single frame of it but to the entire videotape. Consider, for example, a world made up of pointlike particles moving around in three-dimensional space. In four-dimensional space-time — the bird perspective — these particle trajectories resemble a tangle of spaghetti. If the frog sees a particle moving with constant velocity, the bird sees a straight strand of uncooked spaghetti. If the frog sees a pair of orbiting particles, the bird sees two spaghetti strands intertwined like a double helix. To the frog, the world is described by Newton's laws of motion and gravitation. To the bird, it is described by the geometry of the pasta — a mathematical structure. The frog itself is merely a thick bundle of pasta, whose highly complex intertwining corresponds to a cluster of particles that store and process information. Our universe is far more complicated than this example, and scientists do not yet know to what, if any, mathematical structure it corresponds.

The Platonic paradigm raises the question of why the universe is the way it is. To an Aristotelian, this is a meaningless question: The universe just is. But a Platonist cannot help but wonder why it could not have been different. If the universe is inherently mathematical, then why was only one of the many mathematical structures singled out to describe a universe? A fundamental asymmetry appears to be built into the very heart of reality.

As a way out of this conundrum, I have suggested that complete mathematical symmetry holds: that all mathematical structures exist physically as well. Every mathematical structure corresponds to a parallel universe. The elements of this multiverse do not reside in the same space but exist outside of space and time. Most of them are probably devoid of observers. This hypothesis can be viewed as a form of radical Platonism, asserting that the mathematical structures in Plato's realm of ideas or the "mindscape" of mathematician Rudy Rucker of San Jose State University exist in a physical sense. It is akin to what cosmologist John D. Barrow of the University of Cambridge refers to as "$\pi$ in the sky," what the late Harvard University philosopher Robert Nozick called the principle of fecundity, and what the late Princeton philosopher David K. Lewis called modal realism. Level IV brings closure to the hierarchy of multiverses, because any self-consistent fundamental physical theory can be phrased as some kind of mathematical structure.

The Level IV multiverse hypothesis makes testable predictions. As with Level II, it involves an ensemble (in this case, the full range of mathematical structures) and selection effects. As mathematicians continue to categorize mathematical structures, they should find that the structure describing our world is the most generic one consistent with our observations. Similarly, our future observations should be the most generic ones that are consistent with our past observations, and our past observations should be the most generic ones that are consistent with our existence.

Quantifying what "generic" means is a severe problem, and this investigation is only now beginning. But one striking and encouraging feature of mathematical structures is that the symmetry and invariance properties that are responsible for the simplicity and orderliness of our universe tend to be generic, more the rule than the exception. Mathematical structures tend to have them by default, and complicated additional axioms must be added to make them go away.

## What Says Occam?

The scientific theories of parallel universes, therefore, form a four-level hierarchy, in which universes become progressively more different from ours. They might have different initial conditions (Level I); different physical constants and particles (Level II); or different physical laws (Level IV). It is ironic that Level III is the one that has drawn the most fire in the past decades, because it is the only one that adds no qualitatively new types of universes.

In the coming decade, dramatically improved cosmological measurements of the microwave background and the large-scale matter distribution will support or refute Level I by further pinning down the curvature and topology of space. These measurements will also probe Level II by testing the theory of chaotic eternal inflation. Progress in both astrophysics and high-energy physics should also clarify the extent to which physical constants are fine-tuned, thereby weakening or strengthening the case for Level II.

If current efforts to build quantum computers succeed, they will provide further evidence for Level III, as they would, in essence, be exploiting the parallelism of the Level III multiverse for parallel computation. Experimenters are also looking for evidence of unitarity violation, which would rule out Level III. Finally, success or failure in the grand challenge of modern physics — unifying general relativity and quantum field theory — will sway opinions on Level IV. Either we will find a mathematical structure that exactly matches our universe, or we will bump up against a limit to the unreasonable effectiveness of mathematics and have to abandon that level.

So should you believe in parallel universes? The principal arguments against them are that they are wasteful and that they are weird. The first argument is that multiverse theories are vulnerable to Occam's razor because they postulate the existence of other worlds that we can never observe. Why should nature be so wasteful and indulge in such opulence as an infinity of different worlds? Yet this argument can be turned around to argue *for* a multiverse. What precisely would nature be wasting? Certainly not space, mass, or atoms — the uncontroversial Level I multiverse already contains an infinite amount of all three, so who cares if nature wastes some more? The real issue here is the apparent reduction in simplicity. A skeptic worries about all the information necessary to specify all those unseen worlds.

But an entire ensemble is often much simpler than one of its members. This principle can be stated more formally using the notion of algorithmic information content. The algorithmic information content in a number is, roughly speaking, the length of the shortest computer program that will produce that number as output. For example, consider the set of all integers. Which is simpler, the whole set or just one number? Naively, you might think that a single number is simpler, but the entire set can be generated by quite a trivial computer program, whereas a single number can be hugely long. Therefore, the whole set is actually simpler.

Similarly, the set of all solutions to Einstein's field equations is simpler than a specific solution. The former is described by a few equations, whereas the latter requires the specification of vast amounts of initial data on some hypersurface. The lesson is that complexity increases when we restrict our attention to one particular element in an ensemble, thereby losing the symmetry and simplicity that were inherent in the totality of all the elements taken together.

In this sense, the higher-level multiverses are simpler. Going from our universe to the Level I multiverse eliminates the need to specify initial conditions, upgrading to Level II eliminates the need to specify physical constants, and the Level IV multiverse eliminates the need to specify anything at all. The opulence of complexity is all in the subjective perceptions of observers — the frog perspective. From the bird perspective, the multiverse could hardly be any simpler.

The complaint about weirdness is aesthetic rather than scientific, and it really makes sense only in the Aristotelian worldview.

Yet what did we expect? When we ask a profound question about the nature of reality, do we not expect an answer that sounds strange? Evolution provided us with intuition for the everyday physics that had survival value for our distant ancestors, so whenever we venture beyond the everyday world, we should expect it to seem bizarre.

A common feature of all four multiverse levels is that the simplest and arguably most elegant theory involves parallel universes by default. To deny the existence of those universes, one needs to complicate the theory by adding experimentally unsupported processes and ad hoc postulates: finite space, wave function collapse, and ontological asymmetry. Our judgment therefore comes down to which we find more wasteful and inelegant: many worlds or many words. Perhaps we will gradually get used to the weird ways of our cosmos and find its strangeness to be part of its charm.

NICHOLAS WADE

# In Click Languages, an Echo of the Tongues of the Ancients

FROM *The New York Times*

DO SOME OF TODAY'S languages still hold a whisper of the ancient mother tongue spoken by the first modern humans? Many linguists say language changes far too fast for that to be possible. But a new genetic study underlines the extreme antiquity of a special group of languages, raising the possibility that their distinctive feature was part of the ancestral human mother tongue.

They are the click languages of southern Africa. About thirty survive, spoken by peoples like the San, traditional hunters and gatherers, and the Khwe, who include hunters and herders.

Each language has a set of four or five click sounds, which are essentially double consonants made by sucking the tongue down from the roof of the mouth. Outside of Africa, the only language known to use clicks is Damin, an extinct aboriginal language in Australia that was taught only to men for initiation rites.

Some of the Bantu-speaking peoples who reached southern Africa from their homeland in western Africa some 2,000 years ago have borrowed certain clicks from the Khwe, one use being to substitute for consonants in taboo words.

There are reasons to assume that the click languages may be very old. One is that the click speakers themselves, particularly a group of hunter-gatherers of the Kalahari, belong to an extremely ancient genetic lineage, according to analysis of their DNA. They are called the Jul'hoansi, with the upright bar indicating a click. ("Jul'hoansi" is pronounced like "ju-twansi" except that the "tw" is a click sound like the "tsk, tsk" of disapproval.)

All human groups are equally old, being descended from the same ancestral population. But geneticists can now place ethnic groups on a family tree of humankind. Groups at the ends of short twigs, the ones that split only recently from earlier populations, are younger, in a genealogical sense, than those at the ends of long branches. Judged by mitochondrial DNA, a genetic element passed down in the female line, the Jul'hoansis' line of descent is so ancient that it goes back close to the very root of the human family tree.

Most of the surviving click speakers live in southern Africa. But two small populations, the Hadzabe and the Sandawe, live near Lake Eyasi in Tanzania, in eastern Africa. Two geneticists from Stanford, Dr. Alec Knight and Dr. Joanna Mountain, recently analyzed the genetics of the Hadzabe to figure out their relationship to their fellow click speakers, the Ju'hoansi.

The Hadzabe, too, have an extremely ancient lineage that also traces back close to the root of the human family tree, the Stanford team reports today in the journal *Current Biology*. But the Hadzabe lineage and that of the Jul'hoansi spring from opposite sides of the root. In other words, the Hadzabe and the Jul'hoansi have been separate peoples since close to the dawn of modern human existence.

The Stanford team compared them with other extremely ancient groups like the Mbuti of Zaire and the Biaka pygmies of Central African Republic and found the divergence between the Hadzabe and the Jul'hoansi might be the oldest known split in the human family tree.

Unless each group independently invented click languages at some later time, that finding implies that click languages were spoken by the very ancient population from which the Hadzabe and the Jul'hoansi descended. "The divergence of those genetic lineages is among the oldest on earth," Dr. Knight said. "So one could certainly make the inference that clicks were present in the mother tongue."

If so, the modern humans who left Africa some 40,000 years ago and populated the rest of the world might have been click speakers who later lost their clicks. Australia, where the Damin click language used to be spoken, is one of the first places outside Africa known to have been reached by modern humans.

But the antiquity of clicks, if they are indeed extremely ancient, raises a serious puzzle. Joseph Greenberg of Stanford University, the great classifier of the world's languages, put all the click languages in a group he called Khoisan. But Sandawe and Hadzane, the language of the Hadzabe, are what linguists call isolates. They are unlike each other and every other known language. Apart from their clicks, they have very little in common even with the other Khoisan languages.

That the Hadzabe and the Jul'hoansi differ as much in their language as in their genetics is a reflection of the same fact. They are extremely ancient, and there has been a long time for both their language and their genetics to diverge. The puzzle is why they should have retained their clicks when everything else in their languages has changed.

Dr. Knight suggested that clicks might have survived because in the savanna, where most click speakers live, the sounds allow hunters to coordinate activity without disturbing prey. Whispered speech that uses just clicks sounds more like branches creaking than human talk. Clicks make up more than 40 percent of the language and suffice for hunters to convey their meanings, Dr. Knight said.

Dr. Anthony Traill, an expert on click languages at the University of Witwatersrand in South Africa, said he did not find the hunting idea very plausible.

"Clicks are acoustically high-impact sounds for mammalian ears," Dr. Traill said, "probably the worst sounds to use if you are trying to conceal your presence."

But he agreed that it was a puzzle to understand why clicks had been retained for so long. He has found that in the ordinary process of language change, certain types of click can be replaced by nonclick consonants, but he has never seen the reverse occur. "It is highly improbable that a fully fledged click system could arise from nonclick precursors," Dr. Traill said.

Because languages change so fast, it is difficult for linguists to measure their age. Indeed, most think that languages more than a few thousand years old can rarely be dated. But if Dr. Traill is right, that clicks can be lost but not reinvented, that implies that clicks may be a very ancient component of language.

Dr. Bonnie Sands, a linguist at Northern Arizona University, said click sounds were not particularly hard to make. All children can make them. Dr. Sands saw no reason why clicks could not have been invented independently many times and, perhaps, lost in all areas of the world except Africa.

"There is nothing to be gained by assuming that clicks must have been invented only once," she said, "or in presuming that certain types of phonological systems are more primordial than others."

Dr. Traill said that although a single click was not difficult, rattling off a whole series is another matter, because they are like double consonants. "Fluent articulation of clicks in running speech is by any measure difficult," he said. "It requires more articulatory work, like taking two stairs at a time."

Given the laziness of the human tongue, why have clicks been retained by click speakers while everything else changed? "That is a major problem," Dr. Traill said. "All the expectations would be that they would have succumbed to the pressures of change that affect all languages. I do not know the answer."

A leading theory to explain the emergence of behaviorally modern humans 50,000 years ago is that some genetic change enabled one group of people to perfect modern speech. The new power of communication, according to an archaeologist, Dr. Richard Klein, made possible the advanced behaviors that begin to be reflected in the archaeological record of the period.

The Stanford team calculated a date of 112,000 years, plus or minus 42,000 years, for the separation of the Hadzabe and Jul'hoansi populations. If this means that modern speech existed that long ago, it does not appear to fit with Dr. Klein's thesis.

But Dr. Knight said the estimate was very approximate and added that he believed the new findings about click language were fully compatible with Dr. Klein's theory. Clicks might have been part of the first fully articulate human language that appeared among some group of early humans 50,000 years ago. Those with the language gene would have out-competed all other groups, so that language became universal in the surviving human population.

That would explain why the metaphorical Adam hit it off with Eve. They just clicked.

NICHOLAS WADE

# A Prolific Genghis Khan, It Seems, Helped People the World

FROM *The New York Times*

A REMARKABLE LIVING legacy of the Mongol empire has been discovered by geneticists in a survey of human populations from the Caucasus to China.

They find that as many as 8 percent of the men dwelling in the confines of the former Mongol empire bear Y chromosomes that seem characteristic of the Mongol ruling house.

If so, some 16 million men, or half a percent of the world's male population, can probably claim descent from Genghis Khan.

The finding seems to be the first proof, on a genetic level, of the occurrence in humans of sexual selection, a form of sex-based natural selection in which a male or female has an unusual number of offspring. This process can greatly influence the genetic makeup of a species, resulting in otherwise puzzling features like the peacock's cumbersome tail.

The survey was conducted by Dr. Chris Tyler-Smith of Oxford University and geneticist colleagues in China, Pakistan, Uzbekistan, and Mongolia. Over ten years they collected blood from sixteen populations that live in and around the former Mongol empire.

In the late thirteenth century the sons of Genghis Khan controlled territory that stretched from the Pacific coast of China to the Caspian Sea, spanning land now held by the Central Asian republics and the northeast corner of Iran.

Dr. Tyler-Smith's team analyzed the DNA of the Y chromosome, a part of the genome that is useful for establishing human lineages, because, like a surname, it is passed down from father to son.

They found that a cluster of Y chromosomes carried a genetic signature showing they were closely related to one another and to a single founder chromosome in the recent past. These signature chromosomes were far more common than would be expected by chance among most of the populations living within the former Mongol empire. But none of the peoples outside the empire carried the chromosomes, except for the Hazara people of Pakistan and Afghanistan, former Mongol soldiers who claim descent from Genghis Khan.

Dr. Tyler-Smith said the signature chromosomes probably belonged to members of the Mongol ruling house. They could have become so common in part because of the rapes that occurred during the Mongol conquest, but more probably because the Mongol khans had access to large numbers of women in the captive territories they ruled for two centuries. An article about the geneticists' findings has been published electronically by the *American Journal of Human Genetics*.

Genghis Khan's sons and heirs ruled over the various khanates in his empire, and may well have used their position to establish large harems, especially if they followed their father's example. David Morgan, a historian of Mongol history at the University of Wisconsin, said Genghis's eldest son, Tushi, had forty sons.

As for Genghis himself, Dr. Morgan cited a passage from 'Ata-Malik Juvaini, a Persian historian who wrote a long treatise on the Mongols in 1260. Juvaini said: "Of the issue of the race and lineage of Chingiz Khan, there are now living in the comfort of wealth and affluence more than 20,000. More than this I will not say . . . lest the readers of this history should accuse the writer of exaggeration and hyperbole and ask how from the loins of one man there could spring in so short a time so great a progeny."

Dr. Morgan said that since Mongol rulers controlled a large area, it was "perfectly plausible" that they should have fathered many children. "It's pretty clear what they were doing when they were not fighting," he said.

The Mongol rulers' apparent assiduity in propagating their genes has surprised even human behavioral ecologists, researchers who seek to explain many aspects of human society in terms of the pursuit of reproductive advantage.

"I think it's astonishing," said Dr. Robin Dunbar of the University of Liverpool, coauthor of a leading textbook of human behavioral ecology. "This is a staggering example of how a very small lineage can have a hugely disproportionate share of the descendant population."

Dr. Dunbar said it was known that in tribes like the Yanomamo of Brazil, men of high status tended to have more children. But the Mongol study was the first to his knowledge to document this on a genetic level. "It's exactly equivalent to elephant seals slogging it out on the beach — a handful of males get all the matings," he said.

The practice may have been common in human history and would explain why so many male lineages have gone extinct, leaving a single survivor. It could also explain why "Adam," the common ancestor of all Y chromosomes, seems to have lived much earlier than "Eve," the common ancestor of all mitochondria, genetic elements passed down through the female line, Dr. Dunbar said. When some individuals have far more children than others, the formula for calculating the time to the common ancestor yields a much earlier date.

Dr. Tyler-Smith and his colleagues estimate that the common ancestor of the signature chromosomes they found in the Mongol empire populations lived in around A.D. 1000, 162 years before the birth of Genghis Khan.

Dr. Morgan said, "I see no reason why the family shouldn't have descended in a straight line" from that time to Genghis Khan.

The geneticists' evidence for linking the cluster of signature chromosomes to Genghis Khan is necessarily indirect. The Mongol ruler was buried secretly and his tomb has not been found, let alone any bodily remains that might still harbor fragments of DNA. But the signature chromosomes are carried by only a fifth of present-day Mongolian men, suggesting they belonged to an elite group, presumably the lineage of Genghis Khan and his sons.

Dr. Tyler-Smith and his colleagues argue they have found a second link to Genghis Khan, through the Hazaras, whose oral tradition holds that some of them are his direct descendants. The fact that the Hazaras carry the signature chromosome confirms their oral tradition of descent from Genghis and suggests he carried the chromosome too, the geneticists say.

But historians find fault with this argument. Dr. Morris Rossabi, a Mongol expert at Columbia University, described the Hazaras' claim to be direct descendants of Genghis Khan as "untenable."

"They are descendants of troops and guards sent by Chinggis to this region, and I would be very suspicious about a genealogy based on their so-called oral traditions," Dr. Rossabi wrote in an e-mail message. (Chinggis is a more correct spelling of the familiar Genghis.)

The name Hazara, from the Persian word for "thousand," suggests a Mongol military formation and the Hazaras do look Mongol, Dr. Morgan said, although unlike some villagers in Afghanistan who still speak archaic Mongol, the Hazaras themselves speak Dari, a form of Persian. Some Hazaras may have been Mongol soldiers but none of the imperial house ruled in Afghanistan, Dr. Morgan said, making it hard to argue that the Hazaras' signature chromosome comes directly from Genghis.

Asked if the Mongol rulers' vigorous propagation of their genes was default human behavior, given the opportunity, Dr. Dunbar laughed and said it was probably an extreme form, and not universal. But it illustrated the keen interest some men have in using their power and status to maximize their reproductive advantages, he said.

*Contributors' Notes*

*Other Notable Science and*
*Nature Writing of 2003*

# Contributors' Notes

**Scott Atran** is a research director at the National Center for Scientific Research in Paris and a research scientist at the University of Michigan Institute for Social Research. His work on the religious roots of suicide terrorism has been featured in numerous publications, including the *New York Times, Agence France-Presse,* and the *Wall Street Journal.* He is the author of *Cognitive Foundations of Natural History: Towards an Anthropology of Science; In Gods We Trust: The Evolutionary Landscape of Religion;* and *Plants of the Petén Itza' Maya.*

**Ronald Bailey** is the science correspondent for the public policy magazine *Reason.* Previously he was a staff writer for *Forbes* and has worked as a television producer for PBS and ABC News. He is the author or editor of four books on environmental science and policy. His new book, *Liberation Biology: The Scientific and Moral Defense of the Biotech Revolution,* will be published by Prometheus Books in early 2005. Bailey is a member of the Society of Environmental Journalists and the American Society for Bioethics and the Humanities. He is also an adjunct scholar at the Competitive Enterprise Institute and the Cato Institute.

**Philip M. Boffey,** an editorial writer at the *New York Times,* formerly served as a reporter, science and health editor, and deputy editorial page editor for that newspaper. Boffey was a member of two reporting teams that won Pulitzer Prizes: the first in 1986 for a series on the "Star Wars" missile defense system, the second in 1987 for coverage of the *Challenger* space shuttle disaster. He has been president of the National Association of Science Writers and is a director of the Council for the Advancement of Science

Writing. He is the author of *The Brain Bank of America: An Inquiry into the Politics of Science,* an investigation of the National Academy of Sciences. Born in East Orange, N.J., he received an A.B., magna cum laude, in history, from Harvard College in 1958.

**Austin Bunn** has worked as a boat carpenter, a game designer for reality television, and a journalist for the *New York Times Magazine, Wired,* the *Advocate,* the *Village Voice,* and other publications.

**Jennet Conant** has worked as a writer and reporter for *Newsweek* and as a contributing editor and writer for *Vanity Fair, Esquire, GQ, Manhattan Inc.,* and *Harper's Bazaar,* among other publications. Her first book, *Tuxedo Park,* was a *New York Times* bestseller. Currently at work on a second book, she contributes pieces to the Arts & Leisure section of the *New York Times.* She lives in New York City with her husband and son.

**Daniel C. Dennett** is University Professor and director of the Center for Cognitive Studies at Tufts University. He is the author of a number of books, including *Consciousness Explained, Darwin's Dangerous Idea,* and *Freedom Evolves,* and more than two hundred scholarly articles.

**Gregg Easterbrook** is a senior editor of the *New Republic,* a contributing editor of the *Atlantic Monthly,* a visiting fellow at the Brookings Institution, and a columnist for NFL.com. He is the author of *The Progress Paradox.*

**Garrett G. Fagan** is an associate professor at Pennsylvania State University and the author of *Bathing in Public in the Roman World.* He is currently editing *Archaeological Fantasies: How Pseudoarchaeology Misrepresents the Past and Misleads the Public.*

**Jeffrey M. Friedman** has received numerous awards and honors, including election to the National Academy of Science, for his research on the molecular mechanisms that regulate body weight. As a member of the Board on Science Education, he helps make recommendations to improve science education in the United States.

**Atul Gawande** is assistant professor of surgery at Harvard Medical School, assistant professor in health policy at Harvard School of Public Health, and a staff writer for *The New Yorker.* He is a general and endocrine surgeon at Brigham and Women's Hospital in Boston. His book *Complications: A Surgeon's Notes on an Imperfect Science* was a finalist for the National Book Award. He and his family live in Newton, Massachusetts.

**Horace Freeland Judson** calls himself a writer by trade, an academic by accident. He has worked as an editor, book reviewer, theater and art critic, foreign correspondent, and social historian. A fellow of the John D. and Catherine T. MacArthur Foundation, he is the author of a number of books, notably *The Eighth Day of Creation.* His articles have appeared in *Harper's Magazine,* the *Journal of the American Medical Association,* the *Lancet, Life, Nature,* the *New England Journal of Medicine,* the *New Republic, The New Yorker,* the *New York Times Book Review,* and *Smithsonian,* among other publications. He is currently working on three books: one on scientific fraud, one on genetic technology, and a collection of essays.

**Geoffrey Nunberg** is a researcher at Stanford University's Center for the Study of Language and Information and a consulting professor of linguistics at Stanford. He is also chair of the Usage Panel of the *American Heritage Dictionary.* He does regular commentaries on language for NPR's *Fresh Air* and the *New York Times* "Week in Review."

**Mike O'Connor** started the Bird Watcher's General Store in Orleans, Massachusetts, in 1983. As far as he knew, it was the first bird-watching specialty store. Seventeen years later he began writing a birding column for the *Cape Codder,* a local weekly newspaper, to answer his customers' many questions. The column has been popular ever since. An archive of "Ask the Bird Folks" columns can be found on the shop's Web site: BirdWatchers GeneralStore.com.

**Peggy Orenstein** is the author of *Flux: Women on Sex, Work, Kids, Love, and Life in a Half-Changed World,* an examination of the politics and psychology of women's life choices from their midtwenties through their midforties, as well as the best-selling *SchoolGirls: Young Women, Self-Esteem, and the Confidence Gap,* an in-depth study of educational inequity and self-image conflicts among teenage girls in two diverse communities. A contributing writer for the *New York Times Magazine,* Orenstein has also written for the *Los Angeles Times, Discover, Elle, USA Today, Mother Jones, Vogue, Glamour, Salon,* and *The New Yorker* and has appeared on *Nightline, Good Morning America, Today,* and NPR's *Morning Edition,* among other programs. She is at work on a memoir, *Waiting for Daisy.*

**Virginia Postrel** is the author of *The Substance of Style: How the Rise of Aesthetic Value Is Remaking Commerce, Culture, and Consciousness* and *The Future and Its Enemies.* She writes an economics column for the *New York Times* business section.

**Jonathan Rauch** is a columnist for the *National Journal* and a correspondent for the *Atlantic Monthly*. His latest book is *Gay Marriage: Why It Is Good for Gays, Good for Straights, and Good for America*. His articles have appeared in the *New Republic,* the *Economist, Harper's Magazine, U.S. News & World Report,* the *New York Times,* the *Wall Street Journal,* the *Washington Post,* and the *Los Angeles Times,* among other publications. Rauch was born and raised in Phoenix, Arizona, and graduated in 1982 from Yale University.

**Chet Raymo** is professor emeritus of physics and astronomy at Stonehill College in Massachusetts. For twenty years he was a weekly science columnist for the *Boston Globe*. He is the author of a dozen books on science and nature, including, most recently, *Climbing Brandon: Science and Faith on Ireland's Holy Mountain*.

**Ron Rosenbaum** is the author, most recently, of *Explaining Hitler,* and of *The Secret Parts of Fortune,* a collection of essays and journalism from the *New York Times Magazine, Harper's Magazine,* the *Atlantic Monthly, The New Yorker,* and other publications. He writes a biweekly column for the *New York Observer* and has edited an anthology called *Those Who Forget the Past: The Question of Anti-Semitism.* He is currently working on a book about scholars and directors of Shakespeare.

**Steve Sailer,** a market researcher turned journalist, is also a film critic for the *American Conservative* and the Monday columnist for VDARE.com. He was the national correspondent for United Press International and has contributed to the *National Interest, National Review, National Post* of Toronto, and many other periodicals, not all of which have the word "National" in the title.

**Robert Sapolsky** is professor of biology and neurology at Stanford University and research associate with the Institute of Primate Research, National Museum of Kenya. A neuroendocrinologist, his research focuses on the effects of stress. Two of his books, *Why Zebras Don't Get Ulcers: An Updated Guide to Stress, Stress-Related Disease, and Coping* and *The Trouble with Testosterone: And Other Essays on the Biology of the Human Predicament* were *Los Angeles Times* Book Club Finalists. He has received numerous honors and awards for his work, including a MacArthur Fellowship and an Alfred P. Sloan Fellowship. He is a regular contributor to *Discover* and *Science*.

**Eric Scigliano** is the author of *Love, War, and Circuses: The Age-Old Relationship Between Elephants and Humans* and *Puget Sound: Sea Between the Mountains*. His articles have appeared in *Discover, Technology Review, Orion, The*

*New Yorker,* the *New York Times,* and other publications. His next book, *Michelangelo's Mountain,* will explore ancient and modern marble cultures and the quarries of Carrara, Italy, where his ancestors worked as stonecutters. Scigliano makes his home in Seattle.

**Meredith F. Small** contributes regularly to *Discover* and *New Scientist* and is a commentator on NPR's *All Things Considered.* Trained as a primate behaviorist, she now writes on anthropology, natural history, and health. She is the author of five books, including *What's Love Got to Do with It: The Evolution of Human Mating; Our Babies, Ourselves; How Biology and Culture Shape the Way We Parent;* and *Kids,* and is currently working on a book about the anthropology of mental health, titled *The Culture of Our Discontent.* She is a professor of anthropology at Cornell University.

**Max Tegmark,** in this particular universe, is an associate professor of physics at MIT and the University of Pennsylvania. His main research area is in cosmology, measuring the properties of our universe using tools such as the cosmic microwave background and galaxy clustering.

**Nicholas Wade** has worked on the news sections of *Nature* and *Science* and, for many years, for the *New York Times,* as an editorial writer, science editor, and now science reporter. He is the author of several books on science.

# Other Notable Science and Nature Writing of 2003

SELECTED BY TIM FOLGER

BILL MCKIBBEN
　Designer Genes. *Orion,* May/June.
DONALD G. MCNEIL JR.
　Are Men Necessary? *The New York Times,* November 11.
STEVE MIRSKY
　Truth in Advertising. *Scientific American,* April.
　Dropping By. *Scientific American,* June.
CHARLES MOORE
　Trashed. *Natural History,* November.
OLIVER MORTON
　Deep Impact. *Wired,* February.
DENNIS OVERBYE
　What Happened Before the Big Bang? *The New York Times,* November 11.
ROBERT L. PITMAN
　Good Whale Hunting. *Natural History,* December/January.
DAVID QUAMMEN
　The Bear Slayer. *The Atlantic Monthly,* July/August.
JONATHAN RAUCH
　Will Frankenfood Save the Planet? *The Atlantic Monthly,* October.
MICHAEL ROSENWALD
　Yesterday, They Would Have Died. *Popular Science,* October.
DAVID J. ROTHMAN AND SHEILA M. ROTHMAN
　The Organ Market. *The New York Review of Books,* October 23.
MICHAEL RUSE
　The Mismeasure of Science. *Natural History,* July/August.
OLIVER SACKS
　The Mind's Eye. *The New Yorker,* July 28.
ARTHUR SALTZMAN
　My Animal Instincts. *Ascent,* Spring.
BARBARA SEAMAN
　A Woman's World. *Orion,* July/August.
MICHAEL SHERMER
　Remember the Six Billion. *Scientific American,* October.
STEVE SILBERMAN
　The Key to Genius. *Wired,* December.
REBECCA SOLNIT
　The Silence of the Lambswool Cardigans. *Orion,* July/August.
DAN STOBER
　No Experience Necessary. *Bulletin of the Atomic Scientists,* March/April.
KRISTIN C. STOEVER
　Waiting for Dad. *The New York Times,* February 9.
RICHARD STONE
　The Hunt for Hot Stuff. *Smithsonian,* March.
BRUCE STUTZ
　Pumphead. *Scientific American,* July.

TERRIE M. WILLIAMS
  Sunbathing Seals of Antarctica. *Natural History,* October.
KAREN WRIGHT
  The First Earthlings. *Discover,* March.
  Physical Chemistry. *Discover,* July.
BARRY YEOMAN
  Can We Trust Research Done with Lab Mice? *Discover,* July.
CARL ZIMMER
  What If There Is Something Going On in There? *The New York Times Magazine,*
    September 28.

# THE B·E·S·T AMERICAN SERIES®

## THE BEST AMERICAN SHORT STORIES® 2004

**Lorrie Moore, guest editor, Katrina Kenison, series editor.** "Story for story, readers can't beat *The Best American Short Stories* series" (*Chicago Tribune*). This year's most beloved short fiction anthology is edited by the critically acclaimed author Lorrie Moore and includes stories by Annie Proulx, Sherman Alexie, Paula Fox, Thomas McGuane, and Alice Munro, among others.

0-618-19735-4 PA $14.00 / 0-618-19734-6 CL $27.50
0-618-30046-5 CASS $26.00 / 0-618-29965-3 CD $30.00

## THE BEST AMERICAN ESSAYS® 2004

**Louis Menand, guest editor, Robert Atwan, series editor.** Since 1986, *The Best American Essays* series has gathered the best nonfiction writing of the year and established itself as the best anthology of its kind. Edited by Louis Menand, author of *The Metaphysical Club* and staff writer for *The New Yorker*, this year's volume features writing by Kathryn Chetkovich, Jonathan Franzen, Kyoko Mori, Cynthia Zarin, and others.

0-618-35709-2 PA $14.00 / 0-618-35706-8 CL $27.50

## THE BEST AMERICAN MYSTERY STORIES™ 2004

**Nelson DeMille, guest editor, Otto Penzler, series editor.** This perennially popular anthology is a favorite of mystery buffs and general readers alike. This year's volume is edited by the best-selling suspense author Nelson DeMille and offers pieces by Stephen King, Joyce Carol Oates, Jonathon King, Jeff Abbott, Scott Wolven, and others.

0-618-32967-6 PA $14.00 / 0-618-32968-4 CL $27.50 / 0-618-49742-0 CD $30.00

## THE BEST AMERICAN SPORTS WRITING™ 2004

**Richard Ben Cramer, guest editor, Glenn Stout, series editor.** This series has garnered wide acclaim for its stellar sports writing and topnotch editors. Now Richard Ben Cramer, the Pulitzer Prize–winning journalist and author of the best-selling *Joe DiMaggio*, continues that tradition with pieces by Ira Berkow, Susan Orlean, William Nack, Charles P. Pierce, Rick Telander, and others.

0-618-25139-1 PA $14.00 / 0-618-25134-0 CL $27.50

## THE BEST AMERICAN TRAVEL WRITING 2004

**Pico Iyer, guest editor, Jason Wilson, series editor.** *The Best American Travel Writing 2004* is edited by Pico Iyer, the author of *Video Night in Kathmandu* and *Sun After*

*Dark*. Giving new life to armchair travel this year are Roger Angell, Joan Didion, John McPhee, Adam Gopnik, and many others.

0-618-34126-9 PA $14.00 / 0-618-34125-0 CL $27.50

## THE BEST AMERICAN SCIENCE AND NATURE WRITING 2004

**Steven Pinker, guest editor, Tim Folger, series editor.** This year's edition promises to be another "eclectic, provocative collection" (*Entertainment Weekly*). Edited by Steven Pinker, author of *The Blank Slate* and *The Language Instinct*, it features work by Gregg Easterbrook, Atul Gawande, Peggy Orenstein, Jonathan Rauch, Chet Raymo, Nicholas Wade, and others.

0-618-24698-3 PA $14.00 / 0-618-24697-5 CL $27.50

## THE BEST AMERICAN RECIPES 2004–2005

**Edited by Fran McCullough and Molly Stevens.** "Give this book to any cook who is looking for the newest, latest recipes and the stories behind them" (*Chicago Tribune*). Offering the very best of what America is cooking, as well as the latest trends, timesaving tips, and techniques, this year's edition includes a foreword by the renowned chef Bobby Flay.

0-618-45506-x CL $26.00

## THE BEST AMERICAN NONREQUIRED READING 2004

**Edited by Dave Eggers, Introduction by Viggo Mortensen.** Edited by the best-selling author Dave Eggers, this genre-busting volume draws the finest, most interesting, and least expected fiction, nonfiction, humor, alternative comics, and more from publications large, small, and on-line. This year's collection features writing by David Sedaris, Daniel Alarcón, David Mamet, Thom Jones, and others.

0-618-34123-4 PA $14.00 / 0-618-34122-6 CL $27.50 / 0-618-49743-9 CD $26.00

## THE BEST AMERICAN SPIRITUAL WRITING 2004

**Edited by Philip Zaleski, Introduction by Jack Miles.** The latest addition to the acclaimed Best American series, *The Best American Spiritual Writing 2004* brings the year's finest writing about faith and spirituality to all readers. With an introduction by the best-selling author Jack Miles, this year's volume represents a wide range of perspectives and features pieces by Robert Coles, Bill McKibben, Oliver Sacks, Pico Iyer, and many others.

0-618-44303-7 PA $14.00 / 0-618-44302-9 CL $27.50

HOUGHTON MIFFLIN COMPANY   www.houghtonmifflinbooks.com